**Distributions and The Boundary Values
of Analytic Functions**

Distributions and the Boundary Values of Analytic Functions

E. J. BELTRAMI

State University of New York at Stony Brook

M. R. WOHLERS

*Research Department, Grumman Aircraft Engineering Corporation,
and adjunct staff member of the Polytechnic Institute of Brooklyn*

Academic Press *New York and London 1966*

ACADEMIC PRESS INC.
111 Fifth Avenue, New York, New York 10003

United Kingdom Edition published by
ACADEMIC PRESS INC. (LONDON) LTD.
Berkeley Square House, London W.1

LIBRARY OF CONGRESS CATALOG CARD NUMBER: 66-26259

PRINTED IN THE UNITED STATES OF AMERICA

To
Irene and Jane

Foreword

There is a large and important literature concerned with the question of how to characterize precisely the boundary values of analytic functions. For example, when the analytic functions are of Hardy Class H^2 in a half plane, then their boundary values are attained in the L_2 norm and, in fact, the analytic function can be reproduced by a Cauchy integral of the boundary value. Other characterizations are possible when the domain of holomorphy is a half plane or, more generally, a tubular region in the space of complex n-vectors C^n, and may depend on the fact that the boundary value is the Fourier transform of a function with support in a half axis (in the L_2 case this result is the one-sided Paley-Wiener theorem). Questions of this type frequently arise in the theory of causal operators in association with scattering and dispersion problems in quantum field theory and network theory (e.g., Bremermann, Oehme, and Taylor [Br2], and Youla, Castriotta, and Carlin [Yo1]). However, more recent work in these areas indicates that a thorough study of operators that admit objects other than L_2 functions pose new difficulties which cannot be resolved within the usual framework. Thus, in quantum-field theory one is led to consider boundary values of functions analytic in a tube of C^n, and having polynomial growth properties at infinity. Such functions occur as the Laplace transforms of objects with support in a cone of R^n (see the book by Streater and Wightman [St1]). Although restricted to functions of a single complex variable the same notion occurs in recent work concerning passive or dissipative operators (Wohlers and Beltrami [Wo1]). The principal idea in both is that the Laplace transform takes on its boundary values in a considerably weaker topology than that of L_2 and, in fact, the boundary values are no longer functions but must be interpreted as functionals of a certain type (distributions). Nonetheless it is important that some of the basic L_2 results carry over to this more general setting. It is one purpose of this monograph to extend many of the ideas associated with the work of Paley and Wiener, and others in harmonic analysis, to a distributional framework and, although we deal exclusively with functions of a single variable, the same essential notions will occur when the holomorphy domains are in C^n. In particular, a class of distributions, the "tempered" distributions of Schwartz, will be rather completely studied with this idea in mind and we will see that a certain class of holomorphic functions in a half plane having polynomial growth at infinity, the class $H^{\cdot}$, are as intimately related to their boundary values as the H^2 functions are to theirs.

The first chapter in the present work is an introduction to certain essential notions in the theory of distributions. Our treatment is not comprehensive, but rather we have attempted to survey those results of Schwartz [Sc1] which are required for application in later chapters. In particular, in order to keep the discussion of topology to a minimum, certain important theorems are stated but not proven; in any case their proof is available in the standard references. On the other hand, some other results are given in detail either because they may not be as readily available in the literature or because they serve to illustrate rather well the methods and techniques of distribution theory. A feature of this chapter is a treatment of the spaces originally introduced by Sobolev and Friedrichs in their study of weak and strong solutions of differential equations.

In the second chapter the basic results concerning Laplace transforms of distributions in a strip and half plane are given, including a treatment of the boundary behavior of such transforms, based on recent work.

The last chapter extends the notion of Cauchy integral to distributions and applies it to a circle of questions concerning those analytic functions which are reproduced by Cauchy integrals and, in particular, to the existence of reciprocal Hilbert transform pairs for their boundary values. Woven throughout this discussion is the crucial idea that certain tempered distributions are Fourier transforms of objects which vanish on a half line. We also look at an extension of the continuation theorem of Painleve [Pa1], and prove a version of the celebrated "edge of the wedge" theorem. Then causal and, in particular, passive or dissipative matrix-valued linear operators, as well as lossless ones, are characterized by a careful study of the distributional boundary behavior of the operator "resolvent." Here one deals with resolvents that are either positive real matrices or bounded real (scattering) matrices. One of the relevant and significant results in this section is a distributional converse to a well-known theorem of Fatou ([Fa1]; see also [B4]). Holomorphic functions of polynomial growth in a half plane (which includes positive and bounded real functions) are of some importance in the study of dispersion and scattering, as well as in certain questions of harmonic analysis, and so the results of this chapter should prove to be of considerable interest to those applied mathematicians who wish to familiarize themselves with the distributional tools which are appropriate for a study of these areas. Since the material of this chapter is for the most part available only in scattered journal articles and since most of it is new, and some unpublished, the specialist in harmonic analysis, quantum field theory, or network theory should also find something of interest here.

In Appendix I we give a proof of an important but not well-known matrix valued extension, due to Youla [Yo2], of the Cauer representation for positive real functions. This theorem is used in our study of passive operators.

In a second appendix we survey briefly the extension of some of the results in Chapter III to funtions of n complex variables and, in particular, of their application to quantum field theory.

Also we sketch here some ideas concerning those linear causal operators which are describable by partial differential equations. In no sense is this appendix a unified treatment of these ideas, and detailed arguments are avoided; its insertion is merely to indicate to the interested reader some of the flavor of recent researches in the application of distributions.

Preface

This work is intended as a research monograph with a twofold aim: (1) to introduce some of the tools and techniques of the discipline of distribution theory to the applied mathematician, and (2) to survey some of the very recent and otherwise relatively inaccessible results concerning the distributional boundary behavior of analytic functions, and their application.

As to prerequisites, it is assumed that the reader has some grasp of abstract analysis although not too much is really needed to appreciate some of the basic results. Our treatment is not self-contained in that occasional reference is made to classical results, without proof; however, full references are given wherever this is done.

Although there is little overlap in the present book with the recent monograph of Bremermann (Br1), he does treat some interesting analytic representations for distributions and their transforms, of the type first considered by Carleman, and as such, his book is a useful complement to the present work. Moreover, the excellent book by Zemanian (Z1) also discusses certain aspects of passive operator theory using distributions.

We want to acknowledge the cooperation of the Research Department of Grumman Aircraft Engineering Corporation in the preparation of this book. In particular we want to thank Mrs. Mildred Sudwischer who so excellently typed our manuscript. Portions of the research on which this book is based was supported by the Air Force Office of Scientific Research under Contract AF49(638)-1512.

E. J. Beltrami
M. R. Wohlers

September, 1966

Contents

Chapter III. **Distributional Boundary Values of Analytic Functions**

CHAPTER I

Distributions

Introduction

We begin with a very informal discussion of what motivates the notion of distribution. Precise definitions and all details will be given in subsequent sections but we feel free to use here whatever notation and terminology seems necessary. The reader may dispense with this introduction if he chooses, without loss of continuity.

Let a_j be complex scalars and suppose $f \in L_2$. Then the linear operator equation

$$Lu = \sum_{j \leq m} a_j D^j u = f \qquad (1.1)$$

certainly has a smooth solution u whenever $f = 0$ (here, as in what follows, D denotes differentiation). Less clear is the fact that when $f \neq 0$ a meaning can be given to calling some $u \in L_2$ a solution to (1.1) even when u is not differentiable. To see how this can be done consider the adjoint operator to L given by

$$L^* = \sum_{j \leq m} (-1)^j a_j D^j. \qquad (1.2)$$

If we let $\langle u, v \rangle$ denote the integral of uv then a repeated integration by parts shows that

$$\langle Lu, \varphi \rangle = \langle u, L^*\varphi \rangle \qquad (1.3)$$

for all $u, \varphi \in C_0^\infty$, where C_0^∞ consists of all infinitely differentiable functions of compact support (see the following section). Now the right-hand side of (1.3) has a meaning even if $u \in L_2$. This suggests that we use the adjoint relationship in (1.3) in a formal manner and *define* $u \in L_2$ to be a "weak" solution to $Lu = f$ whenever $\langle f, \varphi \rangle = \langle u, L^*\varphi \rangle$ for all $\varphi \in C_0^\infty$. In particular, if $L = D$ then f would be called the weak derivative of u. The word weak is used here to contrast with the notion of strong derivative to be defined in Section 1.6.

It will be observed later that the integral $\langle u, \varphi \rangle \doteq \int u\varphi$ defines a continuous linear functional on C_0^∞ for each locally L_2 function u (the topology on C_0^∞ is discussed in Section 1.2). It is important to note that not every such functional on C_0^∞ can be defined in this manner by some $u \in L_2^{\text{loc}}$. To see this, consider the linear and continuous mapping of C_0^∞ to the complex numbers given by $\delta : \varphi \to \varphi(0)$; here we state the fact (without going into details, since we have not yet defined the C_0^∞ topology) that $\varphi_n \to 0$ in the C_0^∞ sense implies $\varphi_n(0) \to 0$. If it is assumed that $\langle \delta, \varphi \rangle = \int \delta\varphi \, dx$ for some $\delta \in L_2^{\text{loc}}$, then one quickly arrives at a contradiction. Simply let $\{\varphi_j\}$ be a uniformly bounded sequence whose integrals tend to zero as their supports shrink to zero, and for which $\varphi_j(0) = 1$ [see (1.8)]. Then $\int \delta\varphi_j \to 0$ as $j \to \infty$ while $1 \equiv \varphi_j(0)$ (again the reader is asked to forego details until Section 1.2). The functional δ is the celebrated Dirac delta.

When $\varphi \in C_0^\infty$ the weak derivative of δ is defined as the functional $D\delta$ on C_0^∞ for which

$$\langle D\delta, \varphi \rangle = -\langle \delta, D\varphi \rangle. \qquad (1.4)$$

in complete analogy to (1.3); notice that the adjoint to D is $-D$. Since $D\varphi$ is in C_0^∞ for all φ the right-hand and hence left-hand side of (1.4) is well defined as a continuous linear functional on C_0^∞ because of the linearity and continuity of δ. In general, for any element u in the dual space of C_0^∞ the lth derivative of u is defined to be the functional $D^l u$ given by the adjoint relationship

$$\langle D^l u, \varphi \rangle = \langle u, (D^l)^*\varphi \rangle = (-1)^l \langle u, D^l\varphi \rangle. \qquad (1.5)$$

The concepts of weak solution and weak derivative had already been established by the late 1930's by Sobolev [So1] and Friedrichs [Fr1]. A decade later Schwartz developed these ideas into a systematic theory called distributions [Sc1]. The reason for the term distribution is the intimate relationship between the physical notion of distribution of mass (in our case mass on the real axis) and certain linear functionals on C_0^∞. For example, a unit mass placed entirely at the origin defines a function H such that if $H(x)$ is the amount of mass located on the portion of the axis from $-\infty$ to x, then $H(x) = 0$ for $x < 0$ and 1 for $x \geq 0$ (the Heaviside step function). For any $\varphi \in C_0^\infty$ one finds that the Stieltjes integral $\int_{-\infty}^{\infty} \varphi \, dH$ has the value $\varphi(0)$ which, by definition, equals $\langle \delta, \varphi \rangle$, and so δ is formally identified with the weak derivative of H, that is, $\langle DH, \varphi \rangle = -\langle H, D\varphi \rangle = -\int_0^\infty D\varphi \, dx = \varphi(0)$ because of the support of φ; hence $\langle DH, \varphi \rangle = \langle \delta, \varphi \rangle$ for all φ, and so DH and δ define the same functional. The elements of the dual of C_0^∞, called $\mathscr{D}'$, are all known as distributions, even though this class contains much more than δ and its derivatives. As we will see, locally L_2 functions can be imbedded in $\mathscr{D}'$; in fact, the same is true for L_p^{loc}, $1 \leq p \leq \infty$. For this reason, $\mathscr{D}'$ may be considered as an enlargement of L_p^{loc}, a fact which allows some authors to call distributions by the name of generalized functions.

If at this point the reader is unconvinced that distributions constitute a meaningful notion, then he should recall the dilemma posed to algebraists of the 15th and 16th centuries when confronted with the necessity of solving the equation $x^2 + 1 = 0$. No real number satisfies this equation. However, it was eventually realized that by introducing "imaginary" quantities a solution to $x^2 + 1 = 0$ could be given, in some sense, even though it was outside the realm of "real" numbers. Similarly, the need to give meaning to the derivative of functions in L_p^{loc} necessitates the introduction of a "weak" derivative as a functional, even though it is not a function.

Should a linear functional on C_0^∞ not be defined by a function in L_p^{loc}, how does one know it? In the same manner that functions are concretely identified by a knowledge of their point-wise behavior on the real axis, so are functionals identified by a knowledge of what they do to each member of C_0^∞. For this reason the elements of C_0^∞ are called testing functions in that $u \in \mathscr{D}'$ is completely characterized once we evaluate the effect that each $\varphi \in C_0^\infty$ has on the scalar $\langle u, \varphi \rangle$. One then has a sort of "Alice looking glass" in which abstract entities u can be viewed as a set of recognizable numbers $\langle u, \varphi \rangle$. Note that the looking glass is two-sided in that by fixing φ and allowing u to vary, one defines a linear functional on $\mathscr{D}'$ or, more precisely, the relation $\langle u, \varphi \rangle$ is bilinear in u, φ. The mirror effect shows itself also in the fact that the smoothness (i.e., infinite differentiability) of C_0^∞ implies the smoothness of $\mathscr{D}'$ because of (1.5).

Now let the class of testing functions be enlarged to C^∞ (infinitely differentiable functions of arbitrary support). A topology is introduced on C^∞ in Section 1.2. The restriction of this topology to the subspace C_0^∞ is weaker than the given topology of C_0^∞, as will be seen. Thus the dual of C^∞, called $\mathscr{E}'$, is smaller than $\mathscr{D}'$. $\mathscr{E}'$ consists of those objects in $\mathscr{D}'$ which, in a sense to be made precise later, have compact support (cf. Section 1.3; in fact, $u \in L_p^{\text{loc}}$ defines a functional on C^∞ only if it has compact support, for otherwise we can always find a $\varphi \in C^\infty$ with sufficient growth at infinity to make the integral $\langle u, \varphi \rangle$ divergent). Thus $C_0^\infty \subset C^\infty$ and $\mathscr{D}' \supset \mathscr{E}'$ and one notices a curious duality: testing functions of compact support mirror distributions of arbitrary support while testing functions of arbitrary support reflect the distributions of compact support.

It was already remarked that the smoothness of the testing functions implies that of the distributions. If instead of C_0^∞ one used C_0^m as test functions then at most m weak derivatives of an element u in the dual space $\mathscr{D}'_m$ can be defined by means of (1.5). Thus the coarseness of the testing functions (i.e., their lack of smoothness) reflects itself in the coarseness of the corresponding duals. Therefore, although the dual of C_0^0 contains δ it does not contain $D\delta$. It will be seen later that $\mathscr{D}'_0$ is identifiable with the class of all complex-valued Borel measures on the line and so, in particular, δ is the measure that is unity for all Borel sets containing the origin and zero otherwise.

Beginning with Section 1.2 we will study distributions more precisely; first by examining some spaces of testing functions in detail, and then by looking at some other special spaces. It is hoped that the above discussion will motivate to some extent what is to follow in this chapter.

1.1. PRELIMINARIES

In our presentation of distributions it will be assumed that certain facts concerning functional analysis, including the essentials of measure and integration on the line, are known to the reader. Since these notions are used repeatedly in different guises we offer the following brief listing of some of the more important such ideas, without proof. For more details, including proofs, consult the book by Yosida [Y1]. This reference is also basic for the material of Section 1.2.

Let $\mathscr{L}$ denote a linear space. A topology on $\mathscr{L}$ is defined by specifying a set of neighborhoods of the origin having certain properties. The standard way of doing this is as follows. Let $\{\rho_\alpha\}$ be a collection of seminorms on $\mathscr{L}$, indexed by α (a seminorm ρ is a functional having all the properties of a norm except that $\rho(\varphi) = 0$ does not imply $\varphi = 0$ for $\varphi \in \mathscr{L}$). Then, as we vary over all α and over all $\varepsilon > 0$, the sets $\{\varphi \in \mathscr{L} \,|\, \rho_\alpha(\varphi) < \varepsilon\}$ defines the neighborhoods of zero. The resulting seminorm topology is said to be a norm topology if it is defined by a single norm $\| \ \ \|$. Finally, a linear space $\mathscr{L}$ which is a union of linear spaces $\mathscr{L}_j$ can be given a topology by requiring that the neighborhoods of zero in $\mathscr{L}$ be those subsets whose intersection with $\mathscr{L}_j$ are neighborhoods of zero in $\mathscr{L}_j$. $\mathscr{L}$ is then said to be the inductive limit of the $\mathscr{L}_j$ spaces. If $\mathscr{L}$ has a seminorm topology then convergence to zero of $\{\varphi_n\}$ in $\mathscr{L}$ means that $\rho_\alpha(\varphi_n) \to 0$ as $n \to \infty$, for all α, and we write $\varphi_n \to 0$. One says that a topology$_1$ on $\mathscr{L}$ is stronger than a topology$_2$ if convergence in the sense of 1 implies convergence in the sense of 2.

A linear functional u on $\mathscr{L}$ is a linear mapping of $\mathscr{L}$ to the field of complex numbers and is denoted by $\langle u, \varphi \rangle$ or $u(\varphi)$ for $\varphi \in \mathscr{L}$. Note that u is the zero functional if and only if (iff) $\langle u, \varphi \rangle = 0$ for all φ. Moreover, u is continuous on $\mathscr{L}$ whenever $\varphi_n \to 0$ implies $\langle u, \varphi_n \rangle \to 0$. In particular, norms and seminorms are continuous with respect to the topologies they induce. On a normed linear space the continuity of u is equivalent to the existence of a positive constant C such that $|\langle u, \varphi \rangle| \leq C\|\varphi\|$ for all φ.

A topologized linear space $\mathscr{L}$ can always be imbedded as a dense subset of a complete space $\bar{\mathscr{L}}$ (complete in the sense that every Cauchy sequence has a limit) and, moreover, every continuous functional on $\mathscr{L}$ can be uniquely extended as a continuous functional on $\bar{\mathscr{L}}$, i.e., there exists a continuous $\bar{u}$ on $\bar{\mathscr{L}}$ such that $\bar{u} = u$ on $\mathscr{L}$. Another basic extension theorem (a version of the Hahn–Banach theorem) tells us that a continuous linear functional defined on a subspace $\mathscr{L}_0$ of a normed linear space $\mathscr{L}$ can be uniquely extended to all of $\mathscr{L}$.

The collection of all continuous linear functionals on a topologized space $\mathscr{L}$ defines a dual space $\mathscr{L}'$. With $\alpha u_1 + \beta u_2$ given by $\langle \alpha u_1 + \beta u_2, \varphi \rangle = \alpha \langle u_1, \varphi \rangle + \beta \langle u_2, \varphi \rangle$ for all $\varphi \in \mathscr{L}$ and all complex scalars α, β, $\mathscr{L}'$ becomes a linear space while a topology in $\mathscr{L}'$, called the strong topology, can be induced on it from the given topology of $\mathscr{L}$. For example, if $\mathscr{L}$ is a normed space then $\|u\|'$, $u \in \mathscr{L}'$, is given as the infimum of the numbers C for which $|\langle u, \varphi \rangle| \leq C\|\varphi\|$. As one verifies, $\| \ \|'$ is a norm on $\mathscr{L}'$ and, with this norm, $\mathscr{L}'$ coincides with its completion.

The scalar $\langle u, \varphi \rangle$ is, as we saw, a bilinear relation between $u \in \mathscr{L}'$ and $\varphi \in \mathscr{L}$. Moreover, for fixed φ, it is continuous on $\mathscr{L}'$ (in a normed space this follows from $|\langle u, \varphi \rangle| \leq \|u\|'\|\varphi\|$). Therefore every $\varphi \in \mathscr{L}$ defines an element of the dual space $\mathscr{L}''$ of $\mathscr{L}'$, and so $\mathscr{L}$ may be considered as a subspace of $\mathscr{L}''$. If $\mathscr{L} = \mathscr{L}''$ then $\mathscr{L}$ is said to be reflexive. In addition to the strong topology on $\mathscr{L}'$ one can introduce a weaker topology and, in this case, $u_n \to 0$ means that $\langle u_n, \varphi \rangle \to 0$ for all $\varphi \in \mathscr{L}$. Strong implies weak convergence; in the case of a normed space one sees this immediately from the relation $|\langle u, \varphi \rangle| \leq \|u\|'\|\varphi\|$.

Now the stronger the topology on $\mathscr{L}$ the smaller will be its closure or completion but, at the same time, the larger will be its dual. To see this for a normed space note that $\| \ \|_1$ is stronger than $\| \ \|_2$ if $\| \ \|_2 \leq M\| \ \|_1$ for some positive constant M. Then $|\langle u, \varphi \rangle| \leq C\|\varphi\|_2 \leq CM\|\varphi\|_1$, and a little reflection shows the assertions about closure and duals to be true.

Let C^m denote the collection of all complex valued m times continuously differentiable functions on the real axis. By the support of $\varphi \in C^m$, written supp φ, we mean the complement of the maximal open set in which $\varphi = 0$. C_0^m will then denote the subspace of C^m consisting of functions having compact support. C_0^∞ will denote the linear space of all infinitely differentiable functions of compact support.

For any $\varphi \in C_0^0$ introduce the norm $\|\varphi\|_p = (\int |\varphi|^p)^{1/p}$ for $1 \le p < \infty$. Then L_p is abstractly defined as the completion of C_0^0 in this norm and L_∞ is then defined as the dual of L_1. The abstract L_p spaces can be concretely identified with those measurable functions whose pth power of their magnitude is Lebesgue integrable ($1 \le p < \infty$) and with measurable functions that are a.e. bounded ($p = \infty$). The notation L_p^{loc} will be used to refer to those functions which are locally L_p, $1 \le p \le \infty$; locally means on every compact subset of the axis. The L_2 space is reflexive and so every continuous linear functional on L_2 can be written as $\langle u, \bar{v} \rangle$ for some $v \in L_2$ as we vary over $u \in L_2$ (the Riesz representation theorem for L_2). Here $\langle u, v \rangle = \int uv \, dx$ and this notation should not be confused with the inner product notation $(u, v) = \int u\bar{v} \, dx$ for L_2. Thus $(u, v) = \langle u, \bar{v} \rangle$ for $u, v \in L_2$.

Finally, to complete our review, we recall that a Borel measure μ is a complex valued set function which is defined and finite for all bounded Borel subsets of the real line and having the property of countable additivity. Note that $\int_{-\infty}^{\infty} d\mu$ need not be finite although the integral $\langle \mu, \varphi \rangle = \int \varphi \, d\mu$ is well defined for all $\varphi \in C_0^0$.

1.2. Some Spaces of Testing Functions

We have already introduced C_0^m and C_0^∞ (in this chapter all functions are complex valued and are defined on the real axis). If the subscript 0 is missing as in C^m and C^∞, then the functions have arbitrary support. The standard example of something in C_0^∞ is given by

$$\psi(x) = \begin{cases} \lambda \exp\left(-1/(1 - x^2)\right), & |x| \le 1 \\ 0, & |x| > 1, \end{cases} \tag{1.6}$$

where the constant λ is chosen so that

$$\int_{|x|\leq 1} \psi\, dx = 1.$$

ψ is certainly infinitely differentiable for $|x| \neq 1$ while the lth derivative $D^l\psi(x) \to 0$ as $|x| \to 1$. Hence $\psi \in C_0^\infty$. Now we define for future reference ψ_j for $j \geq 1$ by

$$\psi_j(x) = j\psi(jx). \tag{1.7}$$

Then $\psi_j \geq 0$, $\int \psi_j\, dx = 1$, and the support of ψ_j is the interval $|x| \leq 1/j$. We will also need the sequence $\dot\psi_j \in C_0^\infty$ given by

$$\dot\psi_j(x) = \psi(jx). \tag{1.8}$$

In this case $\int \dot\psi_j\, dx \to 0$ as $j \to \infty$, and $|\dot\psi_j(x)| \leq 1$ (indeed, this sequence was already used in the Introduction).

One defines $C_0^m(K)$ as that subspace of C_0^m consisting of functions with support contained in the compact set K. C_0^m is just the union of the $C_0^m(K)$ spaces as we vary over K. A topology on $C_0^m(K)$ is given by the norm

$$\|\varphi\|_{m,K} = \sum_{l\leq m} \max_K |D^l\varphi(x)|. \tag{1.9}$$

An equivalent topology on $C_0^m(K)$ is introduced via the norm

$$\max_{\substack{l\leq m\\ x\in K}} |D^l\varphi(x)| \tag{1.9'}$$

in the sense that convergence in either of the norms (1.9) or (1.9′) implies convergence in the other norm. Similarly, $C_0^\infty(K)$ is defined for each compact K and a topology is induced on it by means of the norms or seminorms (1.9) for all $m \geq 0$. Convergence to zero of $\{\varphi_n\}$ in $C_0^\infty(K)$ thus means that $\|\varphi_n\|_{m,K} \to 0$ for all m.

C_0^∞ can be topologized so as to be the inductive limit of the $C_0^\infty(K)$ spaces by using the fact that C_0^∞ is the union of the $C_0^\infty(K)$ as K varies over increasingly large sets. We admit without proof the fact that a sequence $\{\varphi_n\}$ converges to zero in the C_0^∞ topology iff

(i) supp $\varphi_n \subset K$, for some fixed compact K
(ii) $\varphi_n \to 0$ in the $C_0^\infty(K)$ topology   (1.10)

(see Yosida [Y1] for a proof). *The linear space C_0^∞ with the convergence of (1.10) is denoted by $\mathscr{D}$.* In the same manner a topology on C_0^m is given inductively from that of the $C_0^m(K)$ spaces and so the convergence to zero of $\{\varphi_n\}$ in C_0^m means that

(i) $\operatorname{supp}\varphi_n \subset K$,   for some fixed K

(ii) $\|\varphi_n\|_{m,K} \to 0$. $\qquad\qquad\qquad\qquad\qquad\qquad$ (1.11)

Finally, a topology on C^∞ is defined by means of the seminorms

$$\rho_{m,K}(\varphi) = \sum_{l \le m} \max_K |D^l\varphi(x)|, \qquad (1.12)$$

where K ranges over all compact sets and m over all nonnegative integers. *The linear space C^∞ with this topology is called $\mathscr{E}$* and convergence in $\mathscr{E}$ is clearly equivalent to uniform convergence on all compact sets and in all derivatives.

For any compact K let $u \in L_p$ with $\operatorname{supp} u = K$. The convolution of u with the C_0^∞ functions ψ_j of (1.7) is defined by

$$u * \psi_j = \int u(\zeta)\psi_j(x - \zeta)\,d\zeta, \qquad (1.13)$$

or, equivalently, by

$$\int u\left(x - \frac{\zeta}{j}\right)\psi(\zeta)\,d\zeta \equiv j \int u(\zeta)\psi(j(x - \zeta))\,d\zeta.$$

Lemma 1.1. $u * \psi_j \in C_0^\infty$. Moreover, if Ω is an arbitrary neighborhood of K then $\operatorname{supp} u * \psi_j \subset \Omega$ for j sufficiently large. In addition, if $u = 1$ on K and if $K' \subset K$ then, for sufficiently large j, $u * \psi_j = 1$ on the compact K'.

Proof. $\operatorname{supp}\psi_j(x - \zeta)$ is the set $|x - \zeta| \le 1/j$ which shows that $\operatorname{supp} u * \psi_j \subset \operatorname{supp} u + \operatorname{supp} \psi_j \subset \Omega$ for all sufficiently large j. Moreover (1.13) can be differentiated through the integral sign an arbitrary number of times, and so $u * \psi_j \in C_0^\infty$. Finally, $\operatorname{supp}\psi_j(x - \zeta) \subset K$ if j is large and if x varies in the smaller compact set K'. In this case then $u * \psi_j = 1$ on K'.

Lemma 1.2.   Let $\varphi \in C_0^m$. Then there exists a sequence $\{\varphi_n\}$ in C_0^∞ such that $\varphi_n \to \varphi$ in the C_0^m topology. Thus C_0^∞ is dense in C_0^m for each $m \ge 0$.

PROOF.   By Lemma 1.1 $\varphi * \psi_j \in C_0^\infty$. Also

$$|\varphi(x) - \varphi * \psi_j| \leq \int |\varphi(x) - \varphi(\zeta)|\psi_j(x - \zeta)\, d\zeta$$

$$= \int_{|\varphi(x) - \varphi(\zeta)| \leq \varepsilon} + \int_{|\varphi(x) - \varphi(\zeta)| > \varepsilon}$$

The first term on the right-hand side of the inequality is certainly less than ε while the second term is zero for j sufficiently large for the uniform continuity of φ on K implies that $|x - \zeta| > 1/j$ for all large j whenever $|\varphi(x) - \varphi(\zeta)| > \varepsilon$ (note that $\psi_j(x - \zeta) = 0$ for $|x - \zeta| > 1/j$). Thus $\varphi * \psi_j \to \varphi$ uniformly, as does $D^l(\varphi * \psi_j) = D^l\varphi * \psi_j \to D^l\varphi$ for $l \leq m$ while supp $\varphi * \psi_j$ is contained in some fixed compact set. This proves the Lemma.

We already know from Section 1.1 that C_0^0 is dense in L_p, $1 \leq p < \infty$; it now follows from Lemma 1.2 that C_0^∞ is dense in L_p (i.e., in the L_p norm). Another consequence of the lemma, which we leave without proof, is

Corollary 1.1.   If $u \in L_p^{\text{loc}}$ then $\int_K |u - u * \psi_j|^p \to 0$, as $j \to \infty$, for every compact K.

Now if $u \in L_p^{\text{loc}}$ then u defines a linear functional on C_0^∞ by means of the integral

$$\langle u, \varphi \rangle = \int_{-\infty}^{\infty} u\varphi\, dx \tag{1.14}$$

for all $\varphi \in C_0^\infty$. The integral certainly exists since φ has compact support. Moreover,

$$|\langle u, \varphi \rangle| \leq M\|\varphi\|_{m,K} \tag{1.15}$$

for all $\varphi \in C_0^\infty(K)$ and all $m \geq 0$. The constant M depends on K. Thus the restriction of u to each $C_0^\infty(K)$ is continuous, and hence u belongs to the dual space $\mathcal{D}'$. The correspondence between u and the functional it defines is, moreover, unique in the sense that two functions $u, v \in L_p^{\text{loc}}$ define the same functional iff $u = v$ a.e. This follows from

Theorem 1.1.   Let $u \in L_p^{\text{loc}}$. Then $\langle u, \varphi \rangle = 0$ for all $\varphi \in C_0^\infty$ iff $u = 0$ a.e.

PROOF. The "if" part is obvious. To prove the converse let $\psi \in C_0^0$. By Lemma 1.2 there exists a sequence $\varphi_j \in C_0^\infty$, with supp φ_j bounded by the same fixed compact K, such that $\varphi_j \to \psi$ uniformly. Therefore $\langle u, \psi - \varphi_j \rangle \to 0$ as $j \to \infty$; but $\langle u, \varphi_j \rangle = 0$ by hypothesis and so $\langle u, \psi \rangle = 0$. Now let $v \in L_\infty$, with compact support. Since every measurable function is the point-wise limit a.e. of continuous functions there exists a sequence $\{v_j\}$ in C_0^0 such that $v_j \to v$ a.e. Moreover, the v_j can be clearly chosen to be uniformly bounded by $M = \sup |v(x)|$ and where supp v_j are uniformly bounded by the same compact set. We just showed above that $\langle u, v_j \rangle = 0$ for all j and so, using the bounded convergence theorem, $\langle u, v \rangle = 0$. Now if $u(x) = |u(x)| e^{i\omega(x)}$ we let $v(x) = e^{-i\omega(x)}$ for $|x| \le R$ and $v(x) = 0$ otherwise. Then

$$\langle u, v \rangle = \int_{|x| \le R} |u(x)| \, dx = 0$$

and so $u = 0$ a.e. in $|x| \le R$. Since R is arbitrary we are done.

From now on we identify each $v \in L_p^{loc}$ with the unique functional in $\mathscr{D}'$.

Every Borel measure μ also defines a linear functional on C_0^∞ and, in fact, on C_0^0 via the integral

$$\langle \mu, \varphi \rangle = \int_{-\infty}^\infty \varphi \, d\mu. \tag{1.16}$$

Since the variation of μ is locally bounded then

$$|\langle \mu, \varphi \rangle| \le M \|\varphi\|_{0,K} \tag{1.17}$$

for all $\varphi \in C_0^0(K)$ where μ again depends on K. Hence μ defines a linear continuous functional on C_0^0 and so $\mu \in \mathscr{D}_0'$. An important theorem of Riesz states that, conversely, every linear functional on C_0^0 whose restriction to each $C_0^0(K)$ is continuous is given by (1.16) by some μ. Hence $\mathscr{D}_0'$ is identified with the collection of all Borel measures on the line.

1.3. SCHWARTZ DISTRIBUTIONS

To prepare for this section we need two easy results.

Lemma 1.3.   Let K be any compact set and Ω an arbitrary neighborhood of K. Then there exists a $\varphi \in C_0^\infty$ with supp $\varphi \subset \Omega$ such that $\varphi = 1$ on K and $0 \le \varphi \le 1$.

PROOF.   Let $u = 1$ on K' and zero outside, where K' is a compact set satisfying $K \subset K' \subset \Omega$. Now, from Lemma 1.1, $u * \psi_j \in C_0^\infty$ with supp $u * \psi_j \subset \Omega$ and is equal to one on K, for all j sufficiently large. Any such ψ_j satisfies the requirements of Lemma 1.3.

Lemma 1.4. C_0^∞ is dense in C^∞.

PROOF.   Let $\alpha_j \in C_0^\infty$ be equal to one on the compact set $K_j (K_j \subset K_{j+1}$; $\bigcup_{j=1}^\infty K_j$ covers the real axis). Then whenever $\varphi \in C^\infty$ we have $\alpha_j \varphi \in C_0^\infty$ and $\alpha_j \varphi - \varphi = \varphi(\alpha_j - 1) = 0$ on K_j. Therefore, for j sufficiently large,

$$\sum_{l \le m} \max_K |D^l(\varphi - \alpha_j\varphi)| = 0$$

for each fixed m and K. Hence $\alpha_j \varphi \to \varphi$ in the C^∞ topology.

A distribution u has been defined earlier to be a linear functional on C_0^∞ which is continuous with respect to the topology introduced in Section 1.2. This, of course, is equivalent to asserting that u is a linear functional on C_0^∞ whose restriction to each $C_0^\infty(K)$ is continuous. An alternate characterization of $\mathscr{D}'$ is given by

Theorem 1.2.   A linear functional u is a distribution iff for every compact set K there exist constants C, m such that

$$|\langle u, \varphi \rangle| \le C \|\varphi\|_{m,K}, \qquad (1.18)$$

for all $\varphi \in C_0^\infty(K)$.

PROOF.   Let $\varphi_j \in C_0^\infty$ converge to zero. Then (i) $u_j \in C_0^\infty(K)$ for some compact K and (ii) $\|\varphi_j\|_{m,K} \to 0$ for all m. Therefore, by virtue of (1.18), $\langle u, \varphi_j \rangle \to 0$ and so $u \in \mathscr{D}'$. Conversely, suppose $u \in \mathscr{D}'$ and, for some K, (1.18) does not hold, then, for each m, we can find a $\varphi_m \in C_0^\infty(K)$ such that $|\langle u, \varphi_m \rangle| > m \|\varphi_m\|_{m,K}$. Now there exist scalars λ_m for which $\psi_m = \lambda_m \varphi_m$ and $\langle u, \psi_m \rangle = 1$ so that $1 > m \|\psi_m\|_{m,K}$. Therefore, one has a sequence $\{\psi_m\}$ in $C_0^\infty(K)$ such that, for each n,

$$\|\psi_m\|_{n,K} \le \|\psi_m\|_{m,K} \le 1/m$$

for all $m \geq n$, and so $\{\psi_m\}$ converges to zero in the C_0^∞ topology. This contradicts the fact that $u \in \mathscr{D}'$.

Thus $u \in \mathscr{D}'$ iff, for each compact K, u is continuous on $C_0^\infty(K)$ with respect to the $C_0^m(K)$ norm for some m (m depends on K).

$\mathscr{D}'_m$ will denote the subspace of $\mathscr{D}'$ consisting of distributions which are continuous on each $C_0^\infty(K)$ as K varies and m is fixed. This leads to

DEFINITION 1.1. $u \in \mathscr{D}'$ is of finite order if $u \in \mathscr{D}'_m$ for some m. The order of u is the smallest integer for which this is true.

By virtue of Theorem 1.2 and Definition 1.1, u is of order m iff, for every compact K, there exists a constant C such that

$$|\langle u, \varphi \rangle| \leq C \|\varphi\|_{m,K}$$

for all $\varphi \in C_0^\infty(K)$; m is independent of K.

The support of a function is the complement of the maximal open set on which the function vanishes. In a similar fashion we define the support of a distribution. To do this, however, we must know what it means to say that $u \in \mathscr{D}'$ is zero on an open set Ω of the real axis.

DEFINITION 1.2. $u \in \mathscr{D}'$ is zero on an open subset Ω if $\langle u, \varphi \rangle = 0$ for all $\varphi \in C_0^\infty$ with supp $\varphi \subset \Omega$.

DEFINITION 1.3. If $u \in \mathscr{D}'$ then supp u is the complement of the maximal open set on which it is zero (we omit here the proof of the fact that u is zero on the union of open sets if it is zero on each of them; see Schwartz [Sc1]).

In an earlier section a seminorm topology on C^∞ was given which, on C_0^∞, is weaker than the original topology of C_0^∞. Thus every $u \in \mathscr{E}'$ is, as a linear functional on C_0^∞, continuous with respect to the topology of $\mathscr{D}$, and so belongs to $\mathscr{D}'$. We now want to show that $\mathscr{E}'$ is identical with the class of distributions having compact support. To this end first let $u \in \mathscr{D}'$ have compact support K and let $\alpha \in C_0^\infty$ be equal to one on K (Lemma 1.3). For any $\varphi \in C^\infty$ define $\langle u, \varphi \rangle$ as $\langle u, \alpha\varphi \rangle$, which certainly exists since $\alpha\varphi \in C_0^\infty$; moreover, the definition is independent of α, for if β is another such function then supp$(\alpha - \beta)\varphi$

is in the complement of K, and so $\langle u, \alpha\varphi \rangle = \langle u, \beta\varphi \rangle$. Hence $\langle u, \varphi \rangle$ is uniquely determined as a linear functional on C^∞. Moreover, $\langle u, \varphi \rangle$ coincides with the usual value of u on C_0^∞. Now let $\{\varphi_n\}$ in C^∞ converge to zero in the seminorm topology of $\mathscr{E}$. Then

$$\langle u, \varphi_n \rangle = \langle u, \alpha\varphi_n \rangle,$$

and since supp $\alpha\varphi_n$ is contained in some fixed compact set, $\alpha\varphi_n \to 0$ in $\mathscr{D}$; therefore $\langle u, \varphi_n \rangle \to 0$, since $u \in \mathscr{D}'$, and so $u \in \mathscr{E}'$. To prove the converse let $u \in \mathscr{E}'$. If supp u is not compact we can choose a sequence $\{\varphi_n\}$ in C_0^∞ such that supp φ_n is the complement of $\{x \,||x| < n\}$, and for which $\langle u, \varphi_n \rangle = 1$ (as in the proof of Theorem 1.2, the φ_n can be chosen to satisfy $\langle u, \varphi_n \rangle = 1$). But, quite evidently, $\varphi_n \to 0$ in $\mathscr{E}$ and this leads to a contradiction. One last point remains: let $\varphi \in C^\infty$ and evaluate $u(\varphi)$ for any $u \in \mathscr{E}'$ with supp $u = K$. Is this value the same as the one previously defined for any $u \in \mathscr{D}'$ by $\langle u, \varphi \rangle = \langle u, \alpha\varphi \rangle$? The answer is yes, for if we let $\alpha_j\varphi$ be chosen as in Lemma 1.4, where $\varphi \in C^\infty$, then K will be contained in $\{x \,||x| < j\}$, for sufficiently large, and so $\langle u, \varphi \rangle = \langle u, \alpha_j\varphi \rangle$. But, according to this same lemma, $\alpha_j\varphi \to \varphi$ in $\mathscr{E}$, and so $\langle u, \alpha_j\varphi \rangle \to u(\varphi) \equiv \langle u, \varphi \rangle$. All this is summarized in the next theorem.

Theorem 1.3. $\mathscr{E}'$ is identical with the class of distributions having compact support.

We pass now to some topological properties of $\mathscr{D}'$ and $\mathscr{E}'$.

Definition 1.4. A set B of functions $\varphi_\alpha \in C_0^\infty$ is bounded if

(i) supp $\varphi_\alpha \in K$, for some fixed compact K;

(ii) $\sup_\alpha \|\varphi_\alpha\|_{m,K} \le C_m$, where C_m depends only on m.

Condition (ii) tells us that the mth order derivatives of the φ_α are uniformly bounded.

On $\mathscr{D}'$ a topology (the strong topology) is introduced by means of the seminorms

$$\sup_{\varphi \in B} |\langle u, \varphi \rangle| \qquad (1.19)$$

as B varies over all bounded sets in C_0^∞. Thus, in the strong topology, $u_n \to 0$ means that

$$\sup_{\varphi \in B} \langle u_n, \varphi \rangle \to 0.$$

A weak topology on $\mathscr{D}'$ is given by the seminorms

$$|\langle u, \varphi \rangle| \qquad\qquad (1.20)$$

as φ varies over all elements of C_0^∞. Thus $u_n \to 0$ weakly whenever $\langle u_n, \varphi \rangle \to 0$ for all φ. Since (1.20) is bounded by (1.19) it is clear that strong implies weak convergence. Strong and weak boundedness of a set $B' \subset \mathscr{D}'$ is defined in terms of (1.19) and (1.20), respectively. Thus B' is strongly bounded if

$$\sup_{\varphi \in B} |\langle u, \varphi \rangle| \leq M \quad \text{for all} \quad u \in B'$$

and where M depends on B.

In this book we restrict ourselves to using the weak topology of $\mathscr{D}'$, with few exceptions. The following theorems are therefore admitted without proof. In any case, their proof is beyond the scope of our presentation and the reader should consult Volume 1 of the book by Schwartz [Sc1] for details.

Theorem 1.4. Let $\{u_n\}$ be a sequence in $\mathscr{D}'$. If $u_n \to u$ weakly then $u \in \mathscr{D}'$ and $u_n \to u$ strongly.

Theorem 1.5. The dual of $\mathscr{D}'$ with the strong topology is $\mathscr{D}$.

Theorem 1.6. If $\{u_n\}$ in $\mathscr{D}'$ is weakly bounded it is strongly bounded and there exists a subsequence $\{u_{n_j}\}$ such that $u_{n_j} \to u$ strongly (and hence weakly) to some $u \in \mathscr{D}'$.

Theorem 1.4 says that $\mathscr{D}'$ is closed under weak convergence and that weak and strong convergence are equivalent. Theorem 1.5 shows that $\mathscr{D}$ and $\mathscr{D}'$ are mirror images of each other while Theorem 1.6 is a special case of the fact that bounded sets in $\mathscr{D}'$ are relatively compact.

If $\mathscr{D}'$ is replaced by $\mathscr{E}'$ then Theorems 1.4–1.6 remain valid except that now we use the fact that a set $B \subset C^\infty$ is bounded if

$$\sup_{\alpha} \rho_{m,K}(\varphi_\alpha) \leq M_{m,K}.$$

It should be remarked here that Theorem 1.2 can be refined. In Eq. (1.18) it suffices to consider all $\varphi \in B$, where B is any bounded set in C_0^∞. Since a convergent sequence in $\mathscr{D}$ is certainly bounded then the proof of the refined theorem remains exactly the same. Thus a linear functional on C_0^∞ is a distribution iff it is bounded on every bounded set of C_0^∞. This alternate characterization of the elements of $\mathscr{D}'$ is useful in certain arguments.

Whenever $u_n \in L_p^{\mathrm{loc}}$ it is meaningful to ask whether the fact that $u_n \to 0$ weakly implies that $u_n \to 0$ pointwise a.e., and conversely. The answer to both questions is, in general, no. A detailed study of how the two notions of convergence relate for distributions in L_p^{loc} is to be found in [B1].

Earlier we defined the lth derivative of $u \in \mathscr{D}'$ to be the linear functional $D^l u$ for which

$$\langle D^l u, \varphi \rangle = (-1)^l \langle u, D^l \varphi \rangle \tag{1.21}$$

for each $\varphi \in C_0^\infty$. Thus u is infinitely differentiable in this "weak" sense. If $\varphi_n \to 0$ in the topology of $\mathscr{D}$ then, by (1.21), so does $D^l u$. Hence $D^l u$ is a distribution. Moreover, if $u_n \to 0$ in $\mathscr{D}'$ then so does $D^l u_n$, for any l, again by (1.21). We summarize in the following theorem.

Theorem 1.7. $D^l u \in \mathscr{D}'$, for $u \in \mathscr{D}'$, and if $u_n \to u$ in $\mathscr{D}'$ then $D^l u_n \to D^l u$.

Associated with the operation of differentiation is that of finding primitives. $u_1 \in \mathscr{D}'$ is called a primitive of $u \in \mathscr{D}'$ if $D u_1 = u$. Although the proof of the following theorem is not difficult we admit it without proof since it is not essential to our discussion.

Theorem 1.8. Every $u \in \mathscr{D}'$ has an infinite number of primitives. If u_1, u_2 are primitives of u then they differ by at most a constant.

Corollary 1.2. If $D^l u = 0$ for any $u \in \mathscr{D}'$ then u is a polynomial of degree $\leq l - 1$.

PROOF. From Theorem 1.8 the relation $Du = 0$ has the solution $u = $ constant. Now if $D^2 u = 0$ then, by the same remark, $Du = $ constant. However, $u = x$ also satisfies $D^2 u = 0$ and so $u = x + $ constant.

The proof is completed by induction.

We note here that if $u \in C^1$ then the ordinary and distributional derivatives of u coincide as the integration by parts on $\langle Du, \varphi \rangle$ shows, for $\varphi \in C_0^\infty$. A refinement of this result is that if u is absolutely continuous then the a.e. derivative coincides with its weak derivative. To see this note that Du, which exists a.e., is an L_1 function for which $\langle Du, \varphi \rangle = -\langle u, D\varphi \rangle$ after one again integrates by parts.

The multiplication of two distributions is not defined in general but will be in a special case. If u_1, u_2 are two L_p^{loc} functions then $\langle u_1 u_2, \varphi \rangle = \langle u_1, u_2 \varphi \rangle$ for all $\varphi \in C_0^\infty$. The right side of this expression is also well defined for any $u_1 \in \mathscr{D}'_m$, provided that $u_2 \in C^m$ and, especially, is well defined for any $u_1 \in \mathscr{D}'$ if $u_2 \in C^\infty$ since then $u_2 \varphi$ has compact support and is a suitable test function. This leads us to the

DEFINITION 1.5. The product αu of $\alpha \in C^\infty$ and $u \in \mathscr{D}'$ is the distribution defined by $\langle \alpha u, \varphi \rangle = \langle u, \alpha \varphi \rangle$ for all $\varphi \in C_0^\infty$.

Theorem 1.9 (Leibnitz rule). If $u \in \mathscr{D}'$ and $\alpha \in C^\infty$ then

$$D^l(\alpha u) = \sum_{j \leq l} \binom{l}{j} D^j \alpha D^{l-j} u. \tag{1.22}$$

PROOF. For any $\varphi \in C_0^\infty$ one has, using (1.21), $\langle D(\alpha u), \varphi \rangle = -\langle u, \alpha D\varphi \rangle$, while $\langle (D\alpha)u, \varphi \rangle = \langle u, (D\alpha)\varphi \rangle$ and $\langle \alpha(Du), \varphi \rangle = -\langle u, D(\alpha\varphi) \rangle$. But $D(\alpha\varphi) = \alpha D\varphi + (D\alpha)\varphi$ and so $\langle D(\alpha u), \varphi \rangle = \langle \alpha Du + (D\alpha)u, \varphi \rangle$. Repetition of this argument $l - 1$ times yields (1.22).

In the next two chapters a certain class of distributions will play a fundamental role and it is appropriate to introduce them now.

DEFINITION 1.6. $\mathscr{D}'_+$ consists of all distributions with support in the half axis $[0, \infty)$.

Before closing this section we want to look at some important examples of distributions. In the early 1920's Hadamard introduced the notion of finite part (pf) of a divergent integral [H1]. The idea is this: Suppose u is locally L_p except in a neighborhood of the point $x = a$, and consider the integral sum $[\int^{a-\varepsilon} + \int_{a+\varepsilon}]u \, dx$ for $\varepsilon > 0$. If it is possible to extract from this expression a polynomial in $\log \varepsilon$ and

$1/\varepsilon$, which we call $1(\varepsilon)$, and a remainder $F(\varepsilon)$ whose value is finite as $\varepsilon \to 0$, then $\lim_{\varepsilon \to 0} F(\varepsilon)$ is called the finite part of the divergent integral $\int_{-\infty}^{\infty} u \, dx$ and we write it as pf $\int u \, dx$. The part we subtract out, $1(\varepsilon)$, is divergent as $\varepsilon \to 0$. If $\varphi \in C_0^\infty$ then one can similarly consider the finite part of the integral of the product $u\varphi$ and, in fact, this leads us to defining a linear functional pf u on C_0^∞ as follows:

$$\langle \text{pf } u, \varphi \rangle = \text{pf} \int_{-\infty}^{\infty} u\varphi \, dx. \tag{1.23}$$

In all cases of interest pf u is actually continuous on $\mathscr{D}$ (as in the example below) and so it is a distribution. Note that unless the integral in (1.23) is convergent to begin with, in which case its value is obviously the same as its finite value, the distribution pf u is not defined by a function in L_p^{loc}.

Now consider $u = 1/x$. It is not integrable in a neighborhood of the origin. But $\langle \text{pf } 1/x, \varphi \rangle = $ finite part of $\lim_{\varepsilon \to 0} [\int^{-\varepsilon} + \int_\varepsilon][\varphi(x)/x] \, dx$; this last expression, however, certainly exists as a Cauchy principal value (pv) and so

$$\langle \text{pf } 1/x, \varphi \rangle = \text{pv} \int_{-\infty}^{\infty} \frac{\varphi(x)}{x} \, dx. \tag{1.24}$$

To see that pf $1/x$ or, as we sometimes write, pv $1/x$, is actually in $\mathscr{D}'$ it suffices to establish continuity. Suppose that $\{\varphi_n\}$ in C_0^∞ converges to zero. Then supp $\varphi_n \subset K$, for some fixed interval $K = [-R, R]$, and $D^l\varphi_n \to 0$ uniformly for all $l \geq 0$. Moreover, (1.24) can be written as

$$\text{pv} \int_{-R}^{R} \frac{\varphi_n(x) - \varphi_n(0)}{x} \, dx$$

since

$$\varphi_n(0) \, \text{pv} \int_{-R}^{R} \frac{dx}{x} = 0.$$

Also

$$\left| \frac{\varphi_n(x) - \varphi_n(0)}{x} \right| \leq \max_K |D\varphi_n(x)|;$$

hence

$$\left| \text{pv} \int_{-\infty}^{\infty} \frac{\varphi(x)}{x} \, dx \right| \leq 2R \max_{K} |D\varphi_n(x)| \to 0 \qquad \text{as} \quad n \to \infty,$$

as we wished to show. The weak derivatives $D^l \, \text{pv} \, 1/x$ belong to $\mathscr{D}'$ and a direct calculation (see Schwartz, Volume 1 [Sc1]) shows that

$$D^l \, \text{pv} \, 1/x = \text{pf} \, D^l 1/x = (-1)^l l! \, \text{pf} \, 1/x^{l+1}. \tag{1.25}$$

Note, incidentally, that (1.25) gives meaning to pf $1/x^l$ for all $l > 1$.

1.4   DISTRIBUTIONS OF FINITE ORDER

We have seen that if $u \in L_p^{\text{loc}}$ and, in particular, if $u \in C^0$ then $D^l u \in \mathscr{D}'$. We now show that, locally at least, all distributions are finite-order derivatives of function in C^0.

Theorem 1.10.   Let $u \in \mathscr{D}'$ and let Ω be an open set which has compact closure K on the line. Then there exists an integer n for which, on Ω, $u = D^n v$ and where $v \in C^0$.

PROOF.   If $u \in \mathscr{D}'$ then, by (1.18),

$$|\langle u, \varphi \rangle| \leq C \|\varphi\|_{m,K}, \qquad \varphi \in C_0^\infty(K), \tag{1.26}$$

and where C depends on K. By virtue of a repeated application of the mean value estimate one obtains

$$\max_{K} |\varphi(x)| \leq \text{const} \max_{K} |D\varphi(x)|, \tag{1.27}$$

which certainly holds for any $\varphi \in C_0^\infty(K)$, and so there exists some other constant C_1, which again depends on K, such that

$$\|\varphi\|_{m,K} = \sum_{l \leq m} \max |D^l \varphi(x)| \leq C_1 \max_{K} |D^m \varphi(x)|. \tag{1.28}$$

Moreover, since $\varphi(x) = \int^x D\varphi \, dx$, we have $\max_K |\varphi(x)| \leq \int |D\varphi| \, dx$. Therefore, for yet another constant C,

$$|\langle u, \varphi \rangle| \leq C \int_{-\infty}^{\infty} |D^{m+1} \varphi| \, dx. \tag{1.29}$$

Now the collection of all $m + 1$ order derivatives of functions in $C_0^\infty(K)$ is, certainly, a subspace L_0 of $L_1(K)$. Also, the linear functional defined by $D^{m+1}\varphi \to \langle u, \varphi \rangle$ is, by (1.29), continuous on L_0 with respect to the L_1 norm and so, by the Hahn–Banach theorem, it can be extended to all of $L_1(K)$. Since L_∞ is dual to L_1 there exists a $v \in L_\infty$ for which

$$\langle u, \varphi \rangle = (-1)^{m+1} \int_{-\infty}^{\infty} v D^{m+1}\varphi \, dx = \langle D^{m+1}v, \varphi \rangle \qquad (1.30)$$

for all $\varphi \in C_0^\infty(K)$ and so $u = D^{m+1}v$ on Ω. Now let v be zero outside K. Then $v_0 = \int^x v \, dx$ belongs to C^0 and so, finally, $u = D^n v_0$ on Ω, which proves the theorem with $n = m + 2$.

Note that if u is of order m then $u \in \mathcal{D}_m'$ and so Ω in Theorem 1.10 can be taken to be the entire axis for, by (1.18), m is independent of the compact set K we choose. Hence, there exists a $v \in L_\infty^{loc}$ such that $u = D^{m+1}v$. Then, as before, one constructs a $v_0 \in C^0$ so that $u = D^{m+2}v_0$. This result can be further refined by proving the following theorem.

Theorem 1.11. $u \in \mathcal{D}'$ is of order m iff the $m + 2$ primitive of u is a continuous function v having the property that $D^2 v$ is a measure.

PROOF.   Let $u \in \mathcal{D}_m'$. Then, by the same reasoning used in the previous theorem, (1.28) holds on each K except that now m is independent of K. Therefore, the linear map of all m-order derivatives of functions $\varphi \in C_0^\infty(K)$, a subspace L_0 of $C_0^0(K)$, and given by $D^m\varphi \to \langle v, \varphi \rangle$, can be extended to be linear and continuous on $C_0^0(K)$ since the map is continuous on L_0 with respect to the C_0^0 norm from (1.28). By Riesz theorem, quoted in Section 1.2, it then follows that there exists a Borel measure μ such that, for all $\varphi \in C_0^0(K)$,

$$\langle u, \varphi \rangle = (-1)^n \int D^m\varphi \, d\mu. \qquad (1.31)$$

Associated with every Borel μ is a function $v(x)$, locally of bounded variation, defined by

$$v(x) = \int_a^x d\mu \qquad (a \text{ is a point without measure}) \qquad (1.32)$$

and such that $Dv = \mu$: for $\varphi \in C_0^\infty$, $-\int_{-\infty}^{\infty} v D\varphi \, dx = \int_{-\infty}^{\infty} \varphi \, d\mu$. The primitive of v is then some v_0, which is locally absolutely continuous, so that $D^2 v_0 = \mu$. Thus $\langle u, \varphi \rangle = \langle D^{m+2} v_0, \varphi \rangle$ for $\varphi \in C_0^\infty$.

Conversely, if $u = D^m \mu$ then

$$|\langle u, \varphi \rangle| = \left| \int_{-\infty}^{\infty} D^m \varphi \, d\mu \right| \leq C \max_K |D^m \varphi| \leq C \|\varphi\|_{m,K}, \quad (1.33)$$

where C depends on K and $\varphi \in C_0^\infty(K)$. Hence $u \in \mathcal{D}'_m$ which proves the theorem.

The next result characterizes an important class of finite order distributions.

Theorem 1.12. If $u \in \mathscr{E}'$ then $u \in \mathcal{D}'_m$ for some m.

PROOF. Let $u \in \mathscr{E}'$ with supp $u = K_0$. By the remarks in Section 1.3 it suffices to show that the restriction of u to each $C_0^\infty(K)$, K compact, is continuous with respect to the $\| \ \|_{m,K}$ norm, for some fixed m. Let $K \supset K_0$ and suppose $\varphi \in C_0^\infty(K)$. Then $\alpha\varphi \in C_0^\infty(K)$, where $\alpha \in C_0^\infty$ is chosen so that supp $\varphi \subset K$ and $\alpha = 1$ on K_0, and so $\langle u, \alpha\varphi \rangle = \langle u, \varphi \rangle$ for all such φ. By Theorem 1.2, $|\langle u, \varphi \rangle| \leq C \|\alpha\varphi\|_{m,K}$ for some fixed m and C where m does not depend on the K we choose. This shows that $u \in \mathcal{D}'_m$.

The distribution $D^l \delta$ has the origin as sole support, for any $l \geq 0$. We see this for if $\varphi \in C_0^\infty$ is such that supp φ does not contain the origin then $\langle D^l \delta, \varphi \rangle = (-1)^l \langle \delta, D^l \varphi \rangle = 0$. It is an interesting and important observation that, conversely, every $u \in \mathscr{E}'$ having the origin as sole support (point support) is a sum of δ and its derivatives.

Theorem 1.13. If $u \in \mathscr{E}'$ has the origin as point support then

$$u = \sum_{l \leq m} a_l D^l \delta,$$

for some scalars a_l and some integer $m > 0$.

PROOF. First we show that if $u \in \mathscr{E}'$ with supp $u = K_0$ and if u is of order m then $\langle u, \varphi \rangle = 0$ for all $\varphi \in C_0^\infty$ for which $D^l \varphi(x) = 0$ when $x \in K_0$, $l \leq m$. In fact, let K_ε be the set of points whose distance from

K_0 is less than or equal to ε. If $v = 1$ on $K_{2\varepsilon}$ and zero elsewhere then, by Lemma 1.1, $\chi_j = v * \psi_j \in C_0^\infty$ with supp $u * \varphi_j \subset K_{3\varepsilon}$ and $\chi_j = 1$ on K_ε for j large. Also $\langle u, \varphi \rangle = \langle u, \chi_j \varphi \rangle$ and so

$$|\langle u, \varphi \rangle| \le C \sum_{l \le m} \max_{K_{2\varepsilon}} |D^l(\varphi \chi_j)|.$$

Since the derivatives of φ of order $\le m$ vanish at $x = y$, $y \in K_0$, Taylor's formula shows that

$$|\varphi(x)| \le \frac{1}{(m+1)!} \max_{0 < \zeta < 1} |D^{m+1}\varphi(y + \zeta(x - y))|$$

and since $|x - y| \le 3\varepsilon$ for $x \in K_{3\varepsilon}$ and all $y \in K_0$ one obtains

$$\max_{K_{3\varepsilon}} |\varphi(x)| \le \text{const } \varepsilon^{m+1} \tag{1.34}$$

since the derivatives of φ are bounded on $K_{3\varepsilon}$. The same estimate applied to $D^l\varphi$ shows that

$$\max_{K_{3\varepsilon}} |D^l\varphi(x)| \le \text{const } \varepsilon^{m+1-l}, \qquad l \le m. \tag{1.35}$$

Moreover, an estimate on $D^l(v * \psi_j) = D^l\chi_j$ shows that

$$\max_{K_{3\varepsilon}} |D^l\chi_j(x)| \le \text{const } \varepsilon^{-l}, \tag{1.36}$$

and so Leibnitz formula (Theorem 1.9) finally shows, in view of (1.35) and (1.36), that

$$|D^l(\varphi \chi_j)| \le \text{const } \varepsilon, \qquad l \le m. \tag{1.37}$$

Hence $\langle u, \varphi \rangle = 0$ for all such φ. Now suppose K_0 is a single point, viz. the origin $y = 0$. For any $\varphi \in C_0^\infty$ form the Taylor expansion of φ about the origin:

$$\varphi(x) = \sum_{l \le m} \frac{x^l D^l}{l!} \varphi(0) + R(x), \tag{1.38}$$

where the remainder $R(x)$ and the first m derivatives of R vanish when $x = 0$. Hence from what we just showed, $\langle u, R \rangle = 0$, and so

$$\langle u, \varphi \rangle = \sum_{l \le m} a_l D^l \varphi(0) = \left\langle \sum_{l \le m} a_l D^l \delta, \varphi \right\rangle \tag{1.39}$$

with $a_l = \langle u, x^l/l! \rangle$.

An immediate corollary of Theorem 1.13 is that if two distributions differ at most at the origin, i.e., if $\langle u_1 - u_2, \varphi \rangle = 0$ for all $\varphi \in C_0^\infty$ with supp φ not containing the origin, then $u_1 - u_2$ is a sum of delta and its derivatives. Note that if the origin is translated to any other point a then the sum involves $\tau_a \delta$ and its derivatives, where $\tau_a \delta$ is defined by

$$\langle \tau_a \delta, \varphi \rangle = \langle \delta, \tau_{-a} \varphi \rangle, \tag{1.40}$$

and where $\tau_{-a}\varphi(x) = \varphi(x + a)$. Thus we note that although two functions which differ at one point define the same distribution, two distributions which differ at a point are not equivalent.

The next theorem solves the problem of giving meaning to the division u/x^l, $u \in \mathscr{D}'$.

Theorem 1.14. If $v \in \mathscr{D}'$ then the equation $x^l u = v$ has a solution $u \in \mathscr{D}'$ and all such solutions differ at most by

$$\sum_{j \leq l-1} a_j D^j \delta.$$

Since we use only a special version of this theorem in Chapter 3 (with a proof given there involving Fourier transforms) we content ourselves with simply a statement of this general result here.

Corollary 1.3. The equation $x^l u = 0$ has the solution

$$u = \sum_{j \leq l-1} D^j \delta.$$

An interesting consequence of Theorem 1.13 will be useful later. We state it as

Corollary 1.4. Let $u \in \mathscr{D}'$. Then $u = u^+ + u^-$ where $u^+ \in \mathscr{D}'_+$ and $u^- \in \mathscr{D}'_-$.

PROOF. Let $\alpha \in C_0^\infty$ have its support in a neighborhood of the origin with $\alpha = 1$ in some smaller such neighborhood. Then $\alpha u \in \mathscr{E}'$ and so, by Theorem 1.12, $\alpha u = D^l v$, where $v \in C_0^0$. But then $\alpha u = D^l v^+ + D^l v^-$ with supp $v^+ \subset (0, \infty)$ and supp $v^- \subset (-\infty, 0)$. Also $(1 - \alpha)u$ can be written as $u_1 + u_2$ where $u_1 \in \mathscr{D}'_+$, $u_2 \in \mathscr{D}'_-$. Thus $u = \alpha u + (1 - \alpha)u$ is the desired decomposition.

Note that this decomposition is not unique. For example, $\delta \in \mathscr{D}'_+$ and $\mathscr{D}'_-$.

1.5   CONVOLUTION AND REGULARIZATION

If $u \in L_p^{loc}$ and $\varphi \in C_0^\infty$ then the convolution of u with φ is defined by the integral

$$u * \varphi = \int_{-\infty}^{\infty} u(\zeta)\varphi(x - \zeta)\, d\zeta = \int_{-\infty}^{\infty} u(x - \zeta)\varphi(\zeta)\, d\zeta = \varphi * u. \quad (1.41)$$

It is clear that $u * \varphi$ is a function of x. It is important to extend this notion to distributions and so we make the following definition.

DEFINITION 1.7.   If $u \in \mathscr{D}'$ and $\varphi \in C_0^\infty$ we define $u * \varphi$ by

$$u * \varphi = \langle u_\zeta, \varphi(x - \zeta) \rangle \quad (1.42)$$

where u_ζ means that u operates on φ as a function of ζ, for each fixed x.

A basic result concerning convolution is

Theorem 1.15.   If $u \in \mathscr{D}', \varphi \in C_0^\infty$ then $u * \varphi \in C^\infty$ and

$$D^l(u * \varphi) = D^l u * \varphi = u * D^l \varphi, \quad (1.43)$$

$$\operatorname{supp} u * \varphi \subset \operatorname{supp} u + \operatorname{supp} \varphi, \quad (1.44)$$

$$(u * \varphi) * \psi = u * (\varphi * \psi) \qquad \text{for} \qquad \psi \in C_0^\infty. \quad (1.45)$$

PROOF.   By (1.42) $u * \varphi = 0$ unless the support of u has a nonempty intersection with the support of $\varphi(x - \zeta)$ as a function of ζ. Thus $u * \varphi$ is nonzero if $\zeta \in \operatorname{supp} u$ and $x - \zeta \in \operatorname{supp} \varphi$ in which case $x \in \operatorname{supp} u + \operatorname{supp} \varphi$, and this shows (1.44). If $x_n \to x$ then $\varphi(x_n - \zeta) \to \varphi(x - \zeta)$ in $\mathscr{D}$ as a function of ζ and so $u * \varphi$ is continuous in x. Similarly, the difference quotient $(\varphi(x + h - \zeta) - \varphi(x - \zeta))/h \to (D\varphi)(x - \zeta)$ in $\mathscr{D}$, again as a function of ζ, as $h \to 0$. Hence $D(u * \varphi)$ exists and equals $u * D\varphi$. But $\langle u_\zeta, (D\varphi)(x - \zeta) \rangle = -\langle u_\zeta, D_\zeta \varphi(x - \zeta) \rangle = \langle Du_\zeta, \varphi(x - \zeta) \rangle$ and so $u * D\varphi = Du * \varphi$, which establishes (1.43). Since $\varphi, \psi \in C_0^\infty$ then $\operatorname{supp} \varphi * \psi \subset \operatorname{supp} \varphi + \operatorname{supp} \psi$. But the set of points $x + y$, with $x \in \operatorname{supp} \varphi$ and $y \in \operatorname{supp} \psi$, is compact and so both sides of (1.45) have meaning. To show they are equal we refer to the argument given in page 157 of Yosida's work [Y1].

If $\{\varphi_n\}$ in C_0^∞ converge to $\varphi \in C_0^\infty$ in the topology of $\mathscr{D}$ then so does the translated sequence $\{\varphi_n(x - \zeta)\}$ for any fixed x. Hence, supp $\varphi_n(x - \zeta) \subset K_x$ for some compact K_x. If x itself varies in some compact K then, by virtue of (1.42), $u * \varphi_n \to u * \varphi$ uniformly on every such K. Moreover, the same is true for $D^l(u * \varphi_n)$, by Theorem 1.15, and so $u * \varphi_n \to u * \varphi$ in $\mathscr{E}$. Note also that if we translate φ by a quantity h, i.e., if $(\tau_h\varphi)(x) = \varphi(x - h)$, then $u * (\tau_h\varphi) = \tau_h(u * \varphi)$ which, again, is immediate from (1.42). Finally, convolution of $u \in \mathscr{D}'$ with $\varphi \in C_0^\infty$ is, for every u, a linear mapping of C_0^∞ into C^∞ since $u * (\alpha\varphi_1 + \beta\varphi_2) = \alpha(u * \varphi_1) + \beta(u * \varphi_2)$ for scalars α, β. To sum up: convolution with fixed $u \in \mathscr{D}'$ is a linear translation invariant and continuous mapping of C_0^∞ into C^∞. Conversely, we have the following theorem.

Theorem 1.16.   Let U be a linear translation invariant, and continuous mapping of C_0^∞ into C^∞. Then there exists a unique $u \in \mathscr{D}'$ such that $U\varphi = u * \varphi$ for all $\varphi \in C_0^\infty$.

PROOF.   Let $\tilde{\varphi}(x) = \varphi(-x)$. Then the mapping $\tilde{\varphi} \to (U\varphi)(0)$ is linear on C_0^∞ and, by hypothesis, it is continuous. Hence it defines a unique distribution u such that $\langle u, \tilde{\varphi} \rangle = (U\varphi)(0)$. But $\langle u, \tilde{\varphi} \rangle$ also equals $(u * \varphi)(0)$ by (1.42). Now replace φ by $\tau_{-x}\varphi$; since τ_{-x} commutes with U, as well as with convolution, one finally obtains that $(U\varphi)(x) = (u * \varphi)(x)$, which is what we wanted to show.

At this point we are in a position to define convolution between $u_1 \in \mathscr{D}'$ and $u_2 \in \mathscr{E}'$. First of all, $u_2 * \varphi$, for any $\varphi \in C_0^\infty$, is itself in C_0^∞ since supp $u_2 * \varphi \subset$ supp $u_2 +$ supp φ and the latter set is certainly compact. Hence $u_1 * (u_2 * \varphi)$ is a well-defined linear mapping of C_0^∞ into C^∞, by Theorem 1.15, for fixed u_1, u_2. Moreover, as we know, it is translation invariant and continuous, and so there exists a unique $u \in \mathscr{D}'$ such that $u_1 * (u_2 * \varphi) = u * \varphi$. We define $u_1 * u_2$ to be that distribution u and one immediately has $(u_1 * u_2) * \varphi = u_1 * (u_2 * \varphi)$, which extends (1.45). Note that since $\delta * \varphi = \langle \delta_\zeta, \varphi(x - \zeta) \rangle = \varphi(x)$ then, in particular, $u_1 * \delta$ is the distribution defined by $(u_1 * \delta) * \varphi = u_1 * (\delta * \varphi) = u_1 * \varphi$ and so $u_1 * \delta = u_1$ for all $u_1 \in \mathscr{D}'$. We leave without proof certain properties of convolution, although they are not especially difficult to establish.

Theorem 1.17. If $u \in \mathscr{D}'$ and $v \in \mathscr{E}'$ then (i) $v * u$ is defined and $v * u = u * v$. (ii) supp $u * v \subset$ supp $u +$ supp v (and, in particular, if $u, v \in \mathscr{E}'$ then so is $u * v$).

If $u \in \mathscr{E}'$ and $\varphi \in C^\infty$ then $u * \varphi$ is again defined by (1.42) and continues to be in C^∞. Theorem 1.15 remains valid in this case and, in fact, Theorem 1.16 can be extended to consider mappings from C^∞ into itself.

We now want to show that if $u \in \mathscr{D}'$, $v \in \mathscr{E}'$ then $u * v$ can also be defined as that distribution for which

$$\langle u * v, \varphi \rangle = \langle u_\zeta, \langle v_\eta, \varphi(\zeta + \eta) \rangle \rangle \qquad (1.46)$$

for all $\varphi \in C_0^\infty$. To establish (1.46) we note that

$$\langle u * v, \varphi \rangle = ((u * v) * \tilde{\varphi})(0) = (u * (v * \tilde{\varphi}))(0)$$
$$= \langle u_\zeta, (v * \tilde{\varphi})(x - \zeta) \rangle(0).$$

But

$$(v * \tilde{\varphi})(x - \zeta) = \langle v_\eta, \tilde{\varphi}(x - \zeta - \eta) \rangle = \langle v_\eta, \varphi(\zeta + \eta - x) \rangle$$

and so

$$\langle u_\zeta, \langle v_\eta, \varphi(\zeta + \eta - x) \rangle \rangle(0) = \langle u_\zeta, \langle v_\eta, \varphi(\zeta + \eta) \rangle \rangle.$$

Convolution is well defined for pairs of distributions u, v other than $u \in \mathscr{D}'$ and $v \in \mathscr{E}'$ and, in particular, if $u, v \in \mathscr{D}'_+$ (see Definition 1.6). We want to prove this fact which will be crucial to us later on. First one shows that $u * \varphi$ has meaning if $u \in \mathscr{D}'_+$ and $\varphi \in C^\infty$ has the proper support. Specifically, we prove the following lemma.

Lemma 1.5. If $u \in \mathscr{D}'_+$ and if $\varphi \in C^\infty$ has its support on the positive half axis $x \geq 0$ then $u * \varphi$ exists, is in C^∞, and has its support in the same half axis. Moreover, $D^l(u * \varphi) = D^l u * \varphi = u * D^l \varphi$.

PROOF. If $x < 0$ then supp $\varphi(x - \zeta)$ and supp u have an empty intersection while if $x \geq 0$ then the intersection is compact. Thus if $K = $ supp $u \cap$ supp φ and if $\alpha \in C_0^\infty$ is equal to one on K then $\langle \alpha u, \varphi(x - \zeta) \rangle$ certainly exists since $\alpha u \in \mathscr{E}'$ and defines $\langle u, \varphi(x - \zeta) \rangle$

independent of the α chosen (see the proof of Theorem 1.3). Moreover, as we already noted, $\langle u, \varphi(x - \zeta) \rangle = 0$ for $x < 0$. Since $D^l(\alpha u) = \alpha D^l u$ on K one can easily establish the differentiability property on $u * \varphi$.

On the basis of Lemma 1.5 it is possible to define convolution between $u_1, u_2 \in \mathscr{D}'_+$ in precisely the same manner as we did above for $u_1 \in \mathscr{D}'$, $u_2 \in \mathscr{E}'$, for now $u_2 * \varphi$ has its support in the positive half axis for any $\varphi \in C_0^\infty$ (same argument as in Lemma 1.5) and so $u_1 * (u_2 * \varphi)$ exists, and itself has support in the half axis $x \geq 0$. The mapping from C_0^∞ to C^∞ so defined is again a linear translation invariant and continuous, and so there exists a unique $u \in \mathscr{D}'$ for which $u * \varphi = u_1 * (u_2 * \varphi)$. By Lemma 1.5, u is actually in $\mathscr{D}'_+$ and, as before, $u_1 * u_2$ is unambiguously defined to be u.

The next theorem is important for applications and is a distributional variant of Theorem 1.16.

Theorem 1.18. Let U be a linear translation invariant and continuous mapping of $\mathscr{E}'$ into $\mathscr{D}'$. Then $Uv = U\delta * v$ for all $v \in \mathscr{E}'$.

PROOF. Let $\psi \in C_0^\infty$ and, without loss of generality, let supp $\psi \subset [0, 1]$. For any $\varphi \in C^\infty$ form the sum

$$\langle (1/N) \sum_{n=0}^{N} \psi(n/N)\delta_{n/N}, \varphi(x) \rangle$$

$$= \frac{1}{N} \sum \psi(n/N) \langle \tau_{n/N}\delta, \varphi \rangle$$

$$= \frac{1}{N} \sum \psi(n/N)\varphi(n/N), \qquad (1.47)$$

where $\tau_{n/N}\delta$ is defined by $\langle \tau_{n/N}\delta, \varphi \rangle = \langle \delta, \tau_{-n/N}\varphi \rangle$. As $N \to \infty$, (1.47) tends to $\int \psi\varphi \, dx$. Now, for any $u \in \mathscr{D}'$ we note that $u *$ is continuous on $\mathscr{E}'$. To see this let us first suppose that $\{v_n\} \in \mathscr{E}'$ and that $v_n \to v$ weakly, and hence strongly, in $\mathscr{E}'$. Since $v \in \mathscr{E}'$ one can certainly choose supp v_n to be contained in some fixed compact set for, if $\alpha \in C_0^\infty$ is equal to one on supp v, then $\langle \alpha v_n, \psi \rangle \to \langle \alpha v, \psi \rangle = \langle v, \psi \rangle$ for all $\psi \in C^\infty$ and supp $\alpha v_n \subset$ supp α. Now let $\varphi \in C_0^\infty$ and note that $\langle u * v_n, \varphi \rangle = \langle u_\zeta, \langle v_n(\eta), \varphi(\zeta + \eta) \rangle \rangle$, by (1.46). Since supp $v_n \subset K$ for some compact K then supp$\langle v_n(\eta), \varphi(\zeta + \eta)\rangle$ is also compact for as η varies in K, $\varphi(\zeta + \eta)$ varies in some other compact set. Moreover, $\langle v_n(\eta), \varphi(\zeta + \eta)\rangle \to \langle v_\eta, \varphi(\zeta + \eta)\rangle$ uniformly in ζ (the set $\varphi(\zeta + \eta)$ is bounded in $\mathscr{D}$ as ζ varies). The same is true concerning any derivative of this expression and so $\langle v_n(\eta), \varphi(\zeta + \eta)\rangle \to \langle v_\eta, \varphi(\zeta + \eta)\rangle$ in $\mathscr{D}$. But then $\langle u_\zeta, \langle v_n(\eta), \varphi(\zeta + \eta)\rangle\rangle \to \langle u_\zeta, \langle v_\eta, \varphi(\zeta + \eta)\rangle\rangle = \langle u * v, \varphi\rangle$ which establishes the continuity of $u *$. Thus the operator $L = U - U\delta *$ (where $U\delta = u \in \mathscr{D}'$) is itself linear translation invariant and continuous on $\mathscr{E}'$. Moreover, $L(\tau_{n/N}\delta) = \tau_{n/N}L\delta = 0$ and so $0 = L(N^{-1}\Sigma\psi(n/N)\tau_{n/N}\delta) \to L\psi = 0$, which shows that $U\psi = U\delta * \psi$ for all $\psi \in C_0^\infty$. Now let $v \in \mathscr{E}'$. Then, as we will see in Corollary 1.5, there exists a sequence $\{\psi_n\}$ in C_0^∞ such that $v_n \to v$ in $\mathscr{E}'$ and supp $\psi_n \subset K$ for some fixed K. But then, by the continuity of U, $U\psi_n \to Uv$ in $\mathscr{D}'$ while $U\psi_n = U\delta * \psi_n$. However, $U\delta * \psi_n \to U\delta * v$, as we saw above. Hence $Uv = U\delta * v$, which proves the theorem.

We remark here that if U is a linear translation invariant and continuous mapping of $\mathscr{D}'_+$ into itself, then essentially the same argument as used in Theorem 1.18 shows that $Uv = u * v$ for all $v \in \mathscr{D}'_+$ and where u is a unique distribution in $\mathscr{D}'_+$.

It was stated above that convolution may exist for other pairs of distributions than the ones discussed so far. In the next section, in fact, convolution is defined for another class of objects. More generally, if $u, v \in \mathscr{D}'$ and if $\varphi \in C_0^\infty$ is such that the right side of (1.46) has meaning then we can retrace backwards the steps used to establish (1.46) and, using the fact that $u * \varphi \equiv \langle u_\zeta, \varphi(x - \zeta)\rangle$ whenever the latter exists, show that $u * v$ is defined.

It was seen that $u * \delta = u$ for any $u \in \mathscr{D}'$. Thus $D^l u * \varphi = u * D^l\varphi = u * D^l\varphi * \delta = u * D^l\delta * \varphi$ for all $\varphi \in C_0^\infty$, by virtue of (1.43). Hence $u * D^l\delta = D^l u$. We now state without proof that Theorem 1.17 continues to hold for any $u, v \in \mathscr{D}'$ provided $u * v$ is defined. Then $D^l(u * v) = (u * v) * D^l\delta = u * D^l v = D^l u * v$, and we summarize by stating the following theorem.

Theorem 1.19. If $u * v$ is defined for $u, v \in \mathscr{D}'$ then $D^l(u * v) = u * D^l v = D^l u * v$.

In earlier sections it was seen that C_0^∞ is dense in C_0^m as well as in C^∞. We now want to establish the next theorem.

Theorem 1.20. C_0^∞ is dense in $\mathscr{D}'$.

PROOF. Let K_j be an increasing sequence of compact sets which exhausts the real axis. Let $\alpha_j \in C_0^\infty$ be equal to one on K_j and suppose $u_j = (\alpha_j u) * \psi_j$, for any $u \in \mathscr{D}'$, and where ψ_j were defined by (1.7). Then $\langle u_j, \psi \rangle$, for any $\psi \in C_0^\infty$, is equal to $\langle \alpha_j u * \psi_j, \psi \rangle = ((\alpha_j u * \psi_j) * \tilde{\psi})$ $(0) = (\alpha_j u * (\psi_j * \tilde{\psi}))(0) = \langle \alpha_j u, \psi_j * \tilde{\psi} \rangle = \langle u, \alpha_j (\tilde{\psi}_j * \psi) \rangle$. Now $\{\tilde{\psi}_j * \psi\}$ is a sequence in C_0^∞ having their supports contained in a fixed compact K and $D^l(\tilde{\psi}_j * \psi) = \tilde{\psi}_j * D^l \psi$ uniformly on K, by precisely the same argument used in proving Lemma 1.2. Hence $\tilde{\psi}_j * \psi \to \psi$ in $\mathscr{D}$, and the same is true of $\alpha_j(\tilde{\psi}_j * \psi)$ if j is large enough so that $K_j \supset K$. Since u is continuous on C_0^∞ it follows that $\langle u, \alpha_j(\tilde{\psi}_j * \psi) \rangle \to \langle u, \psi \rangle$ and so $u_j \to u$ in $\mathscr{D}'$; it is clear, moreover, that $u_j \in C_0^\infty$.

If $u \in \mathscr{E}'$ then the sequence u_j of Theorem 1.20 can be chosen so that supp $u_j \subset K$ for some fixed K since if $\alpha \in C_0^\infty$ equals 1 on K then $\alpha u_j \to u$ in $\mathscr{E}'$. Thus one has the following corollary.

Corollary 1.5. C_0^∞ is dense in $\mathscr{E}'$ and if $u \in \mathscr{E}'$ then $\{u_j\}$ in C_0^∞ can be chosen to tend to u with supp $u_j \subset K$, for some fixed compact K.

The essential ingredient in proving Theorem 1.20 is in showing that $u_j = \alpha_j u * \psi_j \to u$ in the $\mathscr{D}'$ topology. If one were interested in showing only that C^∞ is dense in $\mathscr{D}'$ then the sequence $\{u * \psi_j\}$ in C^∞ would do the trick. The functions $u * \psi_j$ are called the regularizations or mollifiers of u. It should be noted, in particular, that $\psi_j = \delta * \psi_j \to \delta$ in $\mathscr{D}'$.

Knowledge of the fact that C_0^∞ is dense in any given space allows one to prove certain theorems or establish rules of computation by first proving them on the usually simpler class C_0^∞ and then extending continuously to the closure of C_0^∞ in the given topology of the larger class. An example of this is to be found in the proof of Theorem 1.18.

Before closing this section it will be convenient to prove here the following theorem.

Theorem 1.21. Let C_+^∞ denote the subspace of C_0^∞ consisting of functions whose support is contained in the half axis $x \geq 0$. Then C_+^∞ is dense in $\mathscr{D}'_+$.

Proof. Consider $\tau_{1/j}\psi_j \in C_0^\infty$, where the ψ_j are defined by (1.7). Then for any $u \in \mathscr{D}'_+$ we have $\alpha_j u * \tau_{1/j}\psi_j \in C_+^\infty$ by the same argument used to prove Lemma 1.5 and where the α_j are defined in Lemma 1.4. Moreover, by Theorem 1.20, $\alpha_j u * \tau_{1/j}\psi_j \to u$ in $\mathscr{D}'$.

1.6 Sobolev Spaces

In this section we introduce a class of distributional spaces which have considerable interest and study some of their properties. Sobolev [So1] and Friedrichs [Fr2] were among the early investigators of these spaces and some of the theorems in this section are due to them. A somewhat different approach to the topics discussed below is to be found in the notes by Lax [La1].

We begin by introducing on C_0^∞ the norm

$$\|\varphi\|_{m,2} = \left(\sum_{l \leq m} \int_{-\infty}^{\infty} |D^l \varphi|^2\right)^{1/2}, \qquad m \geq 0, \tag{1.48}$$

and by denoting the completion of C_0^∞ in this norm as $H_{m,2}$ or, as sometimes written, $W^{2,m}$ (the Sobolev spaces). Since $\|\ \|_{0,2}$ is just the L_2 norm, $H_{0,2}$ is simply L_2 itself. If $\varphi \in H_{m,2}$ then there exists a sequence $\{\varphi_n\} \in C_0^\infty$ such that $\|\varphi - \varphi_n\|_{m,2} \to 0$ as $n \to \infty$ (since C_0^∞ is dense in $H_{m,2}$). Now $\|\varphi\|_{m,2} \geq \|D^l\varphi\|_{0,2}$ for any $\varphi \in C_0^\infty$, $l \leq m$, and so $\{D^l\varphi_n\}$ is a Cauchy sequence in the L_2 norm for $l \leq m$. Thus $\varphi_n \to \varphi$ in the L_2 sense, while $D^l\varphi_n$ converge to some element in L_2 which we call the lth order strong or norm derivative of φ and denote it by $D^l\varphi$; hence

$$\|\varphi - \varphi_n\|_{m,2}^2 = \sum_{l \leq m} \|D^l\varphi - D^l\varphi_n\|_{0,2}^2 \to 0. \tag{1.49}$$

(Note that the norm derivative is independent of the approximating sequence chosen.)

If $\varphi \in H_{m,2}$ then, certainly, $\varphi \in L_2$ and, more generally, since $\|\ \|_{m,2} \leq \|\ \|_{m+1,2}$, $H_{m+1,2} \subset H_{m,2}$ for all $m \geq 0$.

Lemma 1.6.   For any $\varphi \in C_0^\infty$,

$$\max_{-\infty < x < \infty} |\varphi(x)| \leq \text{const } \|\varphi\|_{1,2}. \tag{1.50}$$

PROOF.   Let I be an interval of unit length about x. By the mean value theorem, $|\varphi(x_0)| = \int_I |\varphi| \, dx \leq \|\varphi\|_{0,2}$ for some $x_0 \in I$. Now $\varphi(x) - \varphi(x_0) = \int_{x_0}^x D\varphi \, dx$ and so $|\varphi(x)| \leq \|\varphi\|_{0,2} + (x - x_0)^{1/2}(\int_{-\infty}^\infty |D\varphi|^2)^{1/2} \leq \|\varphi\|_{0,2} + \|D\varphi\|_{0,2}$, by the Schwartz inequality. But if $a, b \geq 0$ then $(a + b)^2 = a^2 + b^2 + 2ab$ and, since $2ab = 2(a^2b^2)^{1/2} \leq a^2 + b^2$, then $a + b \leq \sqrt{2}(a^2 + b^2)^{1/2}$ so that

$$|\varphi(x)| \leq \sqrt{2}(\|\varphi\|_{0,2}^2 + \|D\varphi\|_{0,2}^2)^{1/2}, \tag{1.51}$$

which establishes (1.50).

An immediate corollary of Lemma 1.6 is that

$$\max |D^m\varphi| \leq \text{const } \|\varphi\|_{m+1,2}, \qquad m \geq 0, \tag{1.52}$$

for all $\varphi \in C_0^\infty$.

Following Schwartz we define $\mathcal{D}_{L_2}$ to be that subspace of C^∞ such that $\|\varphi\|_{m,2} < \infty$ for all $m \geq 0$. $\mathcal{D}_{L_2}$ is topologized by means of the family of norms or seminorms (1.48) as m varies over all nonnegative integers and convergence to zero of a sequence $\{\varphi_n\}$ in $\mathcal{D}_{L_2}$ is then equivalent to $\|\varphi_n\|_{m,2} \to 0$ for all such m. If $\varphi \in \mathcal{D}_{L_2}$ then, certainly, $\varphi \in H_{m,2}$ and, in fact, the lth norm derivative of φ coincides with the ordinary derivative $D^l\varphi$. Thus $\varphi \in \bigcap_{m\geq 0} H_{m,2}$. Conversely, if $\varphi \in \bigcap_{m\geq 0} H_{m,2}$ then there exists a sequence $\{\varphi_n\}$ in C_0^∞ such that $\|\varphi - \varphi_n\|_{m,2} \to 0$ and so, by (1.52), $D^m\varphi_n$ converge uniformly to $D^m\varphi$ for all m. Hence, $D^m\varphi \in C^0$ for all m, or $\varphi \in C^\infty$. There are two consequences of these remarks which we state as

Theorem 1.22. $\mathcal{D}_{L_2} = \bigcap_{m\geq 0} H_{m,2}$ and C_0^∞ is dense in $\mathcal{D}_{L_2}$.

The collection of all continuous linear functionals on $\mathcal{D}_{L_2}$, the dual of $\mathcal{D}_{L_2}$, is denoted by $\mathcal{D}'_{L_2}$ and is given a weak topology by means of the seminorms

$$|\langle u, \varphi \rangle| \tag{1.53}$$

as φ varies over $\mathscr{D}_{L_2}$, for $u \in \mathscr{D}'_{L_2}$. Since C_0^∞ is dense in $\mathscr{D}_{L_2}$, and since the restriction of $u \in \mathscr{D}'_{L_2}$ to C_0^∞ is continuous in $\mathscr{D}$ (the convergence to zero of $\{\varphi_n\}$ in $\mathscr{D}$ implies $\|\varphi_n\|_{m,2} \to 0$ by virtue of the straight-forward estimate $\|\varphi_n\|_{m,2} \le$ constant $\|\varphi_n\|_{m,K}$ for supp $\varphi_n \subset K$), then $\mathscr{D}'_{L_2} \subset \mathscr{D}'$. We will see below that if $u \in \mathscr{D}'_{L_2}$ then u is of finite order and that $L_2 \subset \mathscr{D}'_{L_2}$.

We now want to prove that the strong L_2 derivative $D^l\varphi$ of $\varphi \in H_{m,2}$ is the same as the lth weak derivative of φ for all $l \le m$. First, however, it should be observed that $H_{m,2}$ is actually a Hilbert space if the inner product $(\varphi, \psi)_m$ of $\varphi, \psi \in H_{m,2}$ is taken to be

$$(\varphi, \psi)_m = \sum_{l \le m} \int_{-\infty}^{\infty} D^l\varphi \overline{D^l\psi} \, dx, \tag{1.54}$$

in which case $\|\varphi\|_{m,2} = (\varphi, \varphi)_m^{1/2}$. If v is the strong lth order L_2 derivative of $\varphi \in H_{m,2}$, $l \le m$, then there is a sequence $\{\varphi_n\}$ in C_0^∞ such that $(\varphi_n, D^l\psi)_0 \to (\varphi, D^l\psi)_0$ for any $\psi \in C_0^\infty$ since $|(\varphi_n - \varphi, D^l\psi)_0| \le$ constant $\|\varphi_n - \varphi\|_{0,2}$. But $(-1)^l(\varphi, D^l\psi)_0 = (-1)^l\langle \varphi, D^l\psi \rangle = \langle D^l\varphi, \overline{\psi} \rangle$, where $D^l\varphi$ is the weak derivative of φ. On the other hand, $(-1)^l(\varphi_n, D^l\psi)_0 = (D^l\varphi_n, \psi)_0 \to (v, \psi)_0 = \langle v, \overline{\psi} \rangle$ and so $v = D^l\varphi$. Hence, if $\varphi \in H_{m,2}$ then the weak derivatives $D^l\varphi \in L_2$ for all $l \le m$. Conversely, if φ and its weak derivatives of order $l \le m$ are in L_2 then there exists $\{\varphi_n\}$ in C_0^∞ such that $D^l\varphi_n \to D^l\varphi$ in L_2 norm so that $D^l\varphi$ is the strong derivative of φ. Thus we have the following theorem

Theorem 1.23. $\varphi \in H_{m,2}$ iff the weak derivatives $D^l\varphi \in L_2$ for $l \le m$.

In analogy with Theorem 1.2 one now has the next theorem.

Theorem 1.24. A linear functional u on $\mathscr{D}_{L_2}$ belongs to $\mathscr{D}'_{L_2}$ iff there exists a constant C and an integer m such that

$$|\langle u, \varphi \rangle| \le C\|\varphi\|_{m,2} \tag{1.55}$$

for all $\varphi \in \mathscr{D}_{L_2}$. Alternatively, $u \in \mathscr{D}'_{L_2}$ iff it is continuous with respect to some $\| \quad \|_{m,2}$ norm, i.e., iff $u \in H'_{m,2}$ for some m.

PROOF. Essentially the same proof as that of Theorem 1.2 is valid here and so it suffices to sketch the main ideas. First of all, if $\varphi_n \to 0$ in $\mathscr{D}_{L_2}$ then, by (1.55), $\langle u, \varphi_n \rangle \to 0$ and so $u \in \mathscr{D}'_{L_2}$. Conversely, let $u \in \mathscr{D}'_{L_2}$ and suppose (1.55) does not hold. Then, for each m, we can find a $\varphi_m \in H_{m,2}$ such that $\langle u, \varphi_m \rangle = 1$ and $\|\varphi_m\|_{m,2} < 1/m$. But then $\|\varphi_m\|_{n,2} \leq 1/m$ for all $m \geq n$ and so $\varphi_m \to 0$ in $\mathscr{D}_{L_2}$ which contradicts the fact that $u \in \mathscr{D}'_{L_2}$.

Thus $u \in \mathscr{D}'_{L_2}$ iff u is in the dual of $H_{m,2}$ for some m so that $\mathscr{D}'_{L_2} = \bigcup_{m \geq 0} H'_{m,2}$. (Note that if $u \in H'_{m,2}$ then u is continuous on $H_{m,2}$ for all $n \geq m$ so that $u \in \mathscr{D}'_{L_2}$.) In particular, since the dual of L_2 is L_2 itself, $L_2 \subset \mathscr{D}'_{L_2}$. Also note that since $H_{m+1,2} \subset H_{m,2}$ then $H'_{m+1,2} \supset H'_{m,2}$ for all m. The question now is to characterize each dual space $H'_{m,2}$. The next theorem does this for us.

Theorem 1.25. $u \in \mathscr{D}'_{L_2}$ iff $u = \Sigma_{l \leq m} D^l u_l$, where $u_l \in L_2$ and m is some integer. (The derivatives, of course, are weak derivatives.)

PROOF. For any $\psi \in \mathscr{D}_{L_2}$, $|\langle D^l u_l, \psi \rangle| = |\langle u_l, D^l \psi \rangle| \leq$ constant $\|D^l \psi\|_{0,2} \leq$ constant $\|\psi\|_{m,2}$ so that $|\langle u, \psi \rangle| \leq$ constant $\|\psi\|_{m,2}$ and so $u \in H'_{m,2}$. Conversely, if $u \in \mathscr{D}'_{L_2}$ then it is continuous on $H_{m,2}$ for some m and, since $H_{m,2}$ is a Hilbert space, the Riesz representation theorem tells us that there exists a $v \in H_{m,2}$ such that

$$\langle u, \psi \rangle = (v, \overline{\psi})_m = \sum_{l \leq m} \int D^l v D^l \psi \, dx \qquad (1.56)$$

for all $\psi \in C_0^\infty$. Thus $\langle u, \psi \rangle = \langle \Sigma_{l \leq m} (-1)^l D^{2l} v, \psi \rangle$. But $D^{2l} v \in L_2$ for $2l \leq m$, while $D^{2l} v$ is the $2l - m$ derivative of an L_2 function when $2l > m$ so that

$$\sum_{l \leq m} (-1)^l D^{2l} v = \sum_{l \leq m} D^l u_l, \qquad u_l \in L_2.$$

Since $\psi \in C_0^\infty$ is arbitrary, $u = \Sigma_{l \leq m} D^l u_l$.

Thus $u \in H'_{m,2}$ iff it can be written as a finite sum of weak derivatives of L_2 functions, up to order m. Moreover, the estimates $\|\varphi\|_{m,2} \leq$ constant $\|\varphi\|_{m,K}$, for $\varphi \in C_0^\infty(K)$, shows that every $u \in H'_{m,2}$ satisfies (1.18) with m independent of K. Hence, u is of finite order m (recall that

$u \in \mathscr{D}'$ is of order m iff its restriction to each $C_0^\infty(K)$ is continuous with respect to the norm $\| \quad \|_{m,K}$ with m independent of K). Incidentally, it is clear from Theorem 1.25 that all weak derivatives of $u \in \mathscr{D}'_{L_2}$ are again in $\mathscr{D}'_{L_2}$.

At this juncture we briefly introduce some new spaces of distributions and, in particular, define what is meant by bounded distributions. Let $\mathscr{D}_{L_p}$ be that subspace of C^∞ consisting of functions φ for which

$$\|\varphi\|_{m,p} = \left(\sum_{l \leq m} \int_{-\infty}^{\infty} |D^l\varphi|^p \, dx \right)^{1/p} < \infty \tag{1.57}$$

and introduce a topology on $\mathscr{D}_{L_p}$ by means of the seminorms (1.57) for $m \geq 0$. $\dot{\beta}$ will be that subspace of $\mathscr{D}_{L_\infty}$ consisting of functions which vanish at infinity together with their derivatives. Then $\mathscr{D}'_{L_p}$, $1 < p \leq \infty$ will be the duals of $\mathscr{D}_{L_q}$, $p^{-1} + q^{-1} = 1$, while $\mathscr{D}'_{L_1}$ will be the dual of $\dot{\beta}$. In particular, $\mathscr{D}'_{L_\infty}$ is written as β' and its elements are called bounded distributions. The following theorem holds, but we give it without proof.

Theorem 1.26. $u \in \mathscr{D}'_{L_p}$, $1 \leq p \leq \infty$, iff $u = \Sigma_{l \leq m} D^l u_l$ for some m, where $u_l \in L_p$.

One final note concerning $\mathscr{D}'_{L_2}$. In the section on Fourier transforms we will see that the distribution $pv\, 1/x$ introduced earlier is a non-trivial and important example of something in $\mathscr{D}'_{L_2}$ (in fact, in $H'_{1,2}$).

1.7 TESTING FUNCTIONS OF RAPID DECAY AND DISTRIBUTIONS OF SLOW GROWTH

In order to discuss a significant extension of the notion of Fourier transform in the next section, it will be necessary to introduce here another space of testing functions together with its dual. The results of this section are patterned, in part, on those of Section 1.3 and Section 1.6 and so it will be unnecessary to give the detailed proofs which, in most cases, will be very nearly similar to the earlier ones.

DEFINITION 1.8. S will denote that subspace of C^∞ consisting of functions φ for which

$$\rho_{m,l}(\varphi) = \sup_{-\infty < x < \infty} |x^m D^l \varphi| < \infty \qquad (1.58)$$

for all m, l.

In S a topology is introduced by means of the family of seminorms (1.58) so that $\{\varphi_n\}$ in S converges to zero iff $\rho_{m,l}(\varphi_n) \to 0$, as $n \to \infty$, for all m, l.

The fact that $x^m D^l \varphi$ is bounded implies that $D^l \varphi = 0(|x|^{-m})$, for all $m \geq 0$, as $|x| \to \infty$. For this reason functions in S are said to be of rapid decay at infinity or to vanish more rapidly than any polynomial. Thus, for example, $e^{-x^2} \in S$.

Lemma 1.7. C_0^∞ is dense in S.

PROOF. The same idea as in the proof of Lemma 1.4 is valid here. Let K_j be an increasing sequence of compact sets which exhaust the line and let $\alpha_j \in C_0^\infty$ equal one on K_j. Then $\alpha_j \varphi \in C_0^\infty$ for any $\varphi \in S$ and $\alpha_j \varphi - \varphi = 0$ on K_j. Hence, $\sup_x |x^m D^l(\alpha_j \varphi - \varphi)| \to 0$ as $j \to \infty$ so that $\alpha_j \varphi \to \varphi$ in S.

The collection of all continuous linear functionals on S is denoted by S'. Since C_0^∞ is dense in S and since the restriction of $u \in S'$ to C_0^∞ is continuous in $\mathscr{D}$ (certainly convergence to zero of $\{\varphi_n\}$ in $\mathscr{D}$ implies that $\rho_{m,l}(\varphi_n) \to 0$ since supp $\varphi_n \subset K$), then $S' \subset \mathscr{D}'$. It turns out, in fact, that distributions in S' are of finite order.

An example of a distribution in S' is any function $u \in L_p^{loc}$ whose behavior at infinity is $0(|x|^N)$ for some fixed N, for then $\langle u, \varphi \rangle = \int_{-\infty}^\infty u\varphi \, dx < \infty$ for all $\varphi \in S$ since φ is $0(|x|^{-N-2})$ as $|x| \to \infty$. However, if the growth of u at infinity is not polynomial-like, then $\langle u, \varphi \rangle$ cannot be expected to have a finite value. Thus, for example, if $u(x) = e^x$ then it is always possible to find a $\varphi \in S$ such that $\langle u, \varphi \rangle$ diverges. For this reason S' distributions are said to be of slow growth at infinity or to be tempered (tempered, that is, by their mild growth). Another example of something in S' is given by measures μ for which

$\int_{-\infty}^{\infty} d\mu/[(1 + x^2)^{N/2}] < \infty$ for some N and we call such measures tempered.

A weak topology is introduced on S' by means of the seminorms

$$|\langle u, \varphi \rangle| \tag{1.59}$$

as φ varies over S. Then $\{u_n\}$ in S' converges weakly to zero iff $\langle u_n, \varphi \rangle \to 0$ as $n \to \infty$, for all $\varphi \in S$. A set $B \subset S$ is bounded if, for every nonnegative integers m and l, the functions $x^m D^l \varphi$ are uniformly bounded on the real axis as φ varies in B. Then a set $B' \subset S'$ is bounded if $\sup_{\varphi \in B}|\langle u, \varphi \rangle| \le M$ for all $u \in B'$ and over all bounded sets $B \subset S$ (M depends on B); $\{u_n\}$ in S' converges strongly to zero whenever $\sup_{\varphi \in B} \langle u_n, \varphi \rangle \to 0$. S' shares the same essential topological properties that hold for $\mathscr{D}'$ and, in fact, one has the following theorem which is left without proof (see Schwartz [Sc1], for details).

Theorem 1.27. Let $\{u_n\}$ be a sequence in S'. If $u_n \to u$ weakly then $u_n \to u$ strongly and $u \in S'$.

The next theorem is a very useful representation theorem for S' distributions which, again, we state without proof; however, it may be established by the same kind of argument used in the proof of Theorem 1.25 (see, for example, Friedman [F1]).

Theorem 1.28. A distribution u belongs to S' iff for some m, l we can write $u = D^l(1 + x^2)^{m/2}v$, where v is a bounded continuous function on the real line (the derivative is, of course, a weak derivative). Moreover, for some other m, the same representation holds where v can now be chosen as a continuous L_2 function.

A modification of the reasoning used to establish Theorem 1.28 yields the next theorem.

Theorem 1.29. If $\{u_\alpha\}$ in S' vary in some bounded set then, for some fixed $m, l, u = D^l(1 + x^2)^{m/2}v_\alpha$, where $\{v_\alpha\}$ are uniformly bounded continuous functions.

For the proof of Theorem 1.29 we again refer to Friedman [F1].

It should be remarked here that the relation $u = D^l(1 + x^2)^{m/2}v$ clearly shows that $D^k u \in S'$ whenever $u \in S'$, for all k.

In the next section it will be established that if $u \in S' \cap D'_+$ then $u = D^l(1 + x^2)^{m/2}v$, where $v \in L_2(0, \infty)$. This fact will be of some importance to us later, as will be the result that every $u \in S'$ can be written as $x^m v$ for some m, where $v \in \mathscr{D}'_{L_2}$. Both these results will be proven through the use of Fourier transforms. However, it is convenient to establish here the following theorem.

Theorem 1.30. $\mathscr{D}'_{L_2} \subset S'$.

Proof. Let $u \in \mathscr{D}'_{L_2}$. If $\{\varphi_n\}$ in S converge to zero then

$$\int_{-\infty}^{\infty} |D^l \varphi_n|^2 = \int_{-\infty}^{\infty} \frac{|D^l \varphi_n|}{1 + x^2}(1 + x^2)|D^l \varphi_n| \, dx \le C \sup_x (1 + x^2)|D^l \varphi_n| \to 0$$

so that $\{\varphi_n\}$ converge to zero in $\mathscr{D}_{L_2}$ as well and thus $\langle u, \varphi_n \rangle \to 0$. Hence $u \in S'$.

We showed earlier that C_0^∞ is dense in $\mathscr{D}'$. At this point we establish the following theorem.

Theorem 1.31. C_0^∞ (and hence S) is dense in S' (i.e., in the strong S' sense).

Proof. For any $u \in S'$ let $u_j = (\alpha_j u) * \psi_j$ with the same notation as in the proof of Theorem 1.20. Then, as in that proof, we arrive at the relation $\langle u_j, \psi \rangle = \langle u, \alpha_j(\psi * \tilde{\psi}_j) \rangle$ for any $\psi \in S$. But now

$$x^m D^l[\alpha_j(\tilde{\psi}_j * \psi) - \psi] = x^m\{\alpha_j(\tilde{\psi}_j * D^l\psi) - D^l\psi + (\tilde{\psi}_j * \psi)D^l\alpha_j\} \to 0$$

as $j \to \infty$ uniformly on the axis, as simple estimates show, for all m, l. Thus $\alpha_j(\psi * \tilde{\psi}_j) \to \psi$ in S and so $\langle u_j, \psi \rangle \to \langle u, \psi \rangle$ which proves the theorem if we use Theorem 1.27.

We want now to state a theorem, without proof, which shows that convolution can be used to display the behavior at infinity of distributions (see Schwartz [Sc1]).

Theorem 1.32.

   (i) $u \in \mathscr{D}'_{L_p}$ iff $u * \varphi \in \mathscr{D}_{L_p}$ for all $\varphi \in C_0^\infty$.

   (ii) $u \in S'$ iff $(u * \varphi)/(1 + x^2)^{m/2} \in \mathscr{D}_{L_\infty}$ for all $\varphi \in C_0^\infty$ and some integer $m \ge 0$.

Finally, we will also need the following result.

Corollary 1.6.   Let $u \in S'$ and suppose $u = u^+ + u^-$ where $u^+ \in \mathscr{D}'_+$ and $u^- \in \mathscr{D}'_-$. (We saw in Corollary 1.4 that such a decomposition is always possible.) Then u^+, u^- are actually in S'.

PROOF.   From Theorem 1.28, $u = D^l v = D^l v^+ + D^l v^- = g^+ + g^-$, where supp $g^+ \subset (0, \infty)$, supp $g^- \subset (-\infty, 0)$. Therefore, $0 = (g^+ - u^+) + (g^- - u^-) = h^+ + h^-$ and so $\langle h^+, \varphi \rangle = 0$ for all $\varphi \in C^\infty$ such that supp φ does not contain the origin. Hence, from Theorem 1.13, h^+ is a sum of δ and its derivatives, which is in S'. Thus $g^+ - h^+ = u^+ \in S'$. Similarly for h^-.

1.8   FOURIER TRANSFORMS

If $u \in L_1$ then it is a standard fact that the Fourier transform

$$\mathfrak{F}u \equiv \hat{u}(\omega) = \int_{-\infty}^{\infty} e^{-i\omega t} u(t) \, dt \qquad (1.60)$$

exists and is a bounded continuous function. Moreover, if $\hat{u}$ is itself an L_1 function then the following inversion formula holds:

$$\mathfrak{F}^{-1}\hat{u} = u(t) = \frac{1}{2\pi} \int_{-\infty}^{\infty} e^{it\omega} \hat{u}(\omega) \, d\omega. \qquad (1.61)$$

Our reference for these and other standard facts concerning the classical Fourier transform is the book of Titchmarsh [T1]. In this section we wish to extend the notion of Fourier transform to distributions in S'; to do so, however, requires that the classical transform on S be looked at in detail.

Theorem 1.33.   The Fourier transformation $\mathfrak{F}$ is a one to one and bicontinuous mapping of S onto itself in the sense that if $\varphi \in S$ then $\mathfrak{F}\varphi \in S$, $\mathfrak{F}^{-1}\varphi \in S$, and $\mathfrak{F}(\mathfrak{F}^{-1}\varphi) = \mathfrak{F}^{-1}(\mathfrak{F}\varphi) = \varphi$. Moreover, $\varphi_n \to 0$ in S implies $\hat{\varphi}_n = \mathfrak{F}\varphi_n \to 0$ in S.

PROOF.   Let $\varphi \in S$. Then $(-it)^l \varphi(t) \in S$ and so the integral

$$D^l \varphi(\omega) = \int (-it)^l \, e^{-i\omega t} \varphi(t) \, dt \qquad (1.62)$$

exists for all $l \geq 0$ so that $\hat{\varphi} \in C^{\infty}$. Moreover, a k-fold integration by parts of this integral shows that

$$(i\omega)^k D^l \hat{\varphi}(\omega) = \int e^{-i\omega t} D^k((-it)^l \varphi(t)) \, dt \qquad (1.63)$$

which certainly exists since $D^k((-it)^l \varphi)$ is certainly L_1 for all $k \geq 0$. Hence $(i\omega)^k D^l \hat{\varphi}(\omega)$ is continuous and bounded for each ω, l pair chosen and so, by (1.58), $\hat{\varphi} \in S$. Note, in particular, that

$$(i\omega)^k \hat{\varphi}(\omega) = \int e^{-i\omega t} D^k \varphi(t) \, dt. \qquad (1.64)$$

We now estimate the integral (1.63) by first multiplying and dividing by $1 + t^2$ to obtain

$$\sup_{-\infty < \omega < \infty} |\omega^k D^l \hat{\varphi}(\omega)| \leq C \sup_{-\infty < t < \infty} |(1 + t^2) D^k t^l \varphi(t)| \qquad (1.65)$$

where $C = \int dt/(1 + t^2)$. But then if $\{\varphi_n\}$ in S converge to zero in the topology of S the inequality (1.65) shows that $\hat{\varphi}_n \to 0$ in S also; hence the mapping $\mathfrak{F}: \varphi \to \hat{\varphi}$ is continuous. We now want to compute $\mathfrak{F}^{-1}\hat{\varphi}$. To do this let $\psi \in S$ and consider the integral

$$\int \psi(\omega)\hat{\varphi}(\omega) \, e^{it\omega} \, d\omega = \int \psi(\omega) e^{it\omega} \, d\omega \int e^{-i\zeta\omega}\varphi(\zeta) \, d\zeta. \qquad (1.66)$$

By Fubini's theorem, which is valid since $\varphi, \psi \in L_1$, the integral (1.66) equals $\int \varphi(\zeta) \, d\zeta \int \psi(\omega) e^{-i\omega(\zeta-t)} \, d\omega = \int \varphi(\zeta)\hat{\psi}(\zeta - t) \, d\zeta = \int \varphi(\zeta + t)\hat{\psi}(\zeta)d\zeta$. Now let $\psi_n(\omega) = \psi(\omega/n)$ and note that (1.66) now becomes

$$\int \psi(\omega/n)\hat{\varphi}(\omega) \, e^{it\omega} \, d\omega = \int \hat{\psi}(\zeta)\varphi(t + \zeta/n) \, d\zeta. \qquad (1.67)$$

Now $\hat{\varphi}, \hat{\psi}$ as members of S, are certainly in L_1 so φ, ψ are bounded and continuous. The dominated convergence theorem then allows us to pass to the limit in (1.67) as $n \to \infty$ to obtain

$$\psi(0)\int \hat{\varphi}(\omega) \, e^{it\omega} \, d\omega = \varphi(t) \int \hat{\psi}(\zeta) \, d\zeta. \qquad (1.68)$$

At this point let $\psi(t) = \exp(-t^2/2)$; then it is a classical computation to obtain $\hat{\psi}(\zeta) = \sqrt{2\pi} \exp(-\zeta^2/2)$ so that $\int \hat{\psi}(\zeta) \, d\zeta = 2\pi$. But then, from (1.68) one has

$$\frac{1}{2\pi}\int_{-\infty}^{\infty} \hat{\varphi}(\omega)\, e^{it\omega}\, d\omega = \varphi(t), \tag{1.69}$$

which, of course, is the inversion formula. Thus, not only is it true that $\mathfrak{F}^{-1}\varphi$ exists for all $\varphi \in S$ but, by virtue of (1.69), $\mathfrak{F}(\mathfrak{F}^{-1}\varphi) = \mathfrak{F}^{-1}(\mathfrak{F}\varphi) = \varphi$. The one-one property follows from the uniqueness of $\mathfrak{F}$ on S: if $\mathfrak{F}\varphi = 0$ then $\mathfrak{F}^{-1}(\mathfrak{F}\varphi) = \varphi = 0$.

Not every $u \in L_p^{\text{loc}}$ defines a distribution of slow growth as we saw in Section 1.7. However, if $u \in L_p$ then it does define something in S' by $\langle u, \varphi \rangle = \int_{-\infty}^{\infty} u\varphi\, dt$ for all $\varphi \in S$. Moreover, if $u \in L_1$ then

$$\langle \mathfrak{F}u, \varphi \rangle = \int_{-\infty}^{\infty} \varphi(\omega)\, d\omega \int_{-\infty}^{\infty} u(t)\, e^{-it\omega}\, dt, \tag{1.70}$$

and, by Fubini's rule, this equals $\int u(t)\, dt \int e^{-it\omega}\varphi(\omega)\, d\omega = \langle u, \mathfrak{F}\varphi \rangle$. This leads us to the following definition.

DEFINITION 1.9. If $u \in S'$ then the Fourier transform $\mathfrak{F}u = \hat{u}$ is defined by the distribution for which

$$\langle \mathfrak{F}u, \varphi \rangle = \langle u, \mathfrak{F}\varphi \rangle \qquad \text{for all} \quad \varphi \in S. \tag{1.71}$$

Now if $\varphi_n \to 0$ in S then, by Theorem 1.33, $\hat{\varphi}_n \to 0$ in S and, since $u \in S'$, $\langle u, \hat{\varphi}_n \rangle \to 0$. Hence $\langle \hat{u}, \varphi_n \rangle \to 0$ and so $\hat{u} \in S'$. Note that (1.71) reduces to the classical Parseval's relation $\int \hat{u}\varphi\, dt = \int u\hat{\varphi}\, dt$ whenever $u \in L_1$ and so, in this more general setting, we continue to call (1.71) Parseval's relation. There is, moreover, an analog of Parseval's formula

$$\int u\bar{\varphi}\, dt = \frac{1}{2\pi}\int \hat{u}\bar{\hat{\varphi}}\, dt. \tag{1.72}$$

We now note that the inversion formula (1.69) can also be written as

$$\tilde{\varphi}(t) = \varphi(-t) = \frac{1}{2\pi}\int \hat{\varphi}(\omega)\, e^{-it\omega}\, d\omega = \frac{1}{2\pi}\hat{\hat{\varphi}}(t), \tag{1.73}$$

and so if we define $\langle \tilde{u}, \varphi \rangle = \langle u, \tilde{\varphi} \rangle$, for $u \in S'$, then $\langle \hat{\hat{u}}, \varphi \rangle = \langle \hat{u}, \hat{\varphi} \rangle$

$= \langle u, \hat{\varphi} \rangle = 2\pi \langle u, \tilde{\varphi} \rangle = 2\pi \langle \tilde{u}, \varphi \rangle$, using (1.71) and (1.73). Thus $\hat{\hat{u}} = 2\pi \tilde{u}$ for all $u \in S'$, which is the inversion formula on S'. Moreover, if $\{u_n\}$ in S' converge weakly to zero in S' then so does $\{\hat{u}_n\}$ since $\langle \hat{u}_n, \varphi \rangle = \langle u_n, \hat{\varphi} \rangle \to 0$. Finally, if $\hat{u} = 0$ then $\langle \hat{\hat{u}}, \varphi \rangle = 2\pi \langle \tilde{u}, \varphi \rangle = 2\pi \langle u, \tilde{\varphi} \rangle = 0$ for all $\varphi \in S$ and so $u = 0$. We summarize by stating

Theorem 1.34. The Fourier transformation $\mathfrak{F}$ is a one to one bicontinuous mapping of S' onto itself in the sense that if $u \in S'$ then $\mathfrak{F}u \equiv \hat{u} \in S'$, $\mathfrak{F}^{-1}u \in S'$, and $\mathfrak{F}(\mathfrak{F}^{-1}u) = \mathfrak{F}^{-1}(\mathfrak{F}u) = u$. Moreover, $u_n \to 0$ in S' implies $\mathfrak{F}u_n = \hat{u}_n \to 0$.

An important set of computational rules involving the Fourier transform on S' are the differentiation formulas

$$D^l \hat{u} = \mathfrak{F}[(-it)^l u] \tag{1.74}$$

$$(i\omega)^l \hat{u} = \mathfrak{F}(D^l u). \tag{1.75}$$

To establish (1.74) and (1.75) we use (1.62) and (1.64), valid on S, and note that for any $\varphi \in S$, $\langle D^l \hat{u}, \varphi \rangle = (-1)^l \langle \hat{u}, D^l \varphi \rangle = (-1)^l \langle u, \widehat{D^l \varphi} \rangle = (-1)^l \langle u, (it)^l \hat{\varphi} \rangle = \langle (-it)^l u, \hat{\varphi} \rangle = \langle \widehat{(-it)^l u}, \varphi \rangle$ while $\langle (i\omega)^l \hat{u}, \varphi \rangle = \langle u, \widehat{(i\omega)^l \varphi} \rangle = (-1)^l \langle u, D^l \hat{\varphi} \rangle = \langle D^l u, \hat{\varphi} \rangle = \langle \widehat{D^l u}, \varphi \rangle$.

It is convenient to derive here certain basic formulas which will be useful to us later: If $u \in S'$, $\varphi \in S$ then

$$2\pi \langle u, \varphi \rangle = \langle \hat{u}, \hat{\varphi} \rangle = \langle \hat{u}, \tilde{\hat{\varphi}} \rangle \tag{1.76}$$

$$\langle \bar{u}, \varphi \rangle = \langle \overline{\mathfrak{F}\hat{u}}, \varphi \rangle \tag{1.77}$$

Let $\tau_h \delta$ be the translated δ defined by $\langle \tau_h \delta, \varphi \rangle = \langle \delta, \tau_{-h} \varphi \rangle$, where $\tau_{-h} \varphi(t) = \varphi(t + h)$. Then

$$\widehat{\tau_h \delta} = e^{-ih\omega} \tag{1.78}$$

$$\widehat{e^{iht}} = 2\pi \tau_h \delta \tag{1.79}$$

To prove (1.76) we write $\langle u, \varphi \rangle = \langle u, \tilde{\tilde{\varphi}} \rangle = 1/2\pi \langle u, \hat{\tilde{\varphi}} \rangle = 1/2\pi \langle \hat{u}, \hat{\tilde{\varphi}} \rangle$ while (1.77) follows from the fact that $\bar{\varphi} = \mathfrak{F}\bar{\hat{\varphi}}$ for $\varphi \in S$. Now $\langle \widehat{\tau_h \delta}, \varphi \rangle = \langle \delta, \tau_{-h} \hat{\varphi} \rangle$; but $\tau_{-h} \hat{\varphi} = \hat{\varphi}(t + h)$, and so $\langle \widehat{\tau_h \delta}, \varphi \rangle = \hat{\varphi}(h) = \langle e^{-i\omega h}, \varphi \rangle$, which is (1.78). Also $\langle \widehat{e^{iht}}, \varphi \rangle = \langle e^{iht}, \hat{\varphi} \rangle$ which, by (1.78), is $\langle \widehat{\tau_{-h} \delta}, \hat{\varphi} \rangle = 2\pi \langle \delta, \tau_h \tilde{\varphi} \rangle = 2\pi \langle \delta, \tau_{-h} \varphi \rangle = 2\pi \langle \tau_h \delta, \varphi \rangle$.

Note, in particular, that (1.78) and (1.79) give

$$\hat{\delta} = 1 \quad \text{and} \quad \hat{1} = 2\pi\delta \qquad (1.80)$$

If $\varphi \in S$ then the arguments given in Section 1.3 to show that pv $1/\omega \in \mathscr{D}'$ can be modified slightly to show, in fact, that pv $1/\omega \in S'$. As an illustration of the preceding let us calculate $\widehat{\text{pv } 1/\omega}$: for $\varphi \in S$,

$$\langle \widehat{\text{pv } 1/\omega}, \varphi \rangle = \text{pv} \int_{-\infty}^{\infty} \frac{\hat{\varphi}(\omega)}{\omega}\, d\omega$$

$$= \lim_{\varepsilon \to 0} \left\{ \left(\int_{\varepsilon}^{\infty} + \int_{-\infty}^{-\varepsilon} \frac{d\omega}{\omega} \right) \int_{-\infty}^{\infty} e^{-it\omega}\varphi(t)\, dt \right\}$$

$$= \lim_{\varepsilon \to 0} \int_{-\infty}^{\infty} \varphi(t)\, dt \left[\int_{\varepsilon}^{\infty} + \int_{-\infty}^{-\varepsilon} \frac{e^{-it\omega}}{\omega}\, d\omega \right]$$

$$= \lim_{\varepsilon \to 0} - \int_{-\infty}^{\infty} \varphi(t)\, dt \int_{\varepsilon}^{\infty} \frac{2i \sin \omega t}{\omega}\, d\omega.$$

Now $\int_{\varepsilon}^{\infty} (\sin \omega t/\omega)\, d\omega = \pi/2 \,\text{sgn}\, t - \int_{0}^{\varepsilon} (\sin \omega t/\omega)\, d\omega$. But the last integral is bounded in magnitude by εt and, since $\varphi(t) \in S$, we can apply the dominated convergence theorem to conclude that $\langle \widehat{\text{pv } 1/\omega}, \varphi \rangle = -\langle \pi i \,\text{sgn}\, t, \varphi \rangle$. Hence

$$\widehat{\text{pv } \frac{1}{\omega}} = -\pi i \,\text{sgn}\, t, \qquad (1.81)$$

and similarly, $\widehat{\frac{1}{2} \,\text{sgn}\, t} = \text{pv } 1/i\omega$. Now let $H(t)$ be the Heaviside step. Then $H(t) = \frac{1}{2}(1 + \text{sgn}\, t)$ and so $\hat{H} = \frac{\hat{1}}{2} + \frac{1}{2} \widehat{\text{sgn}\, t}$. But, from (1.80), $\frac{\hat{1}}{2} = \pi\delta$ while from (1.81) $\frac{1}{2} \widehat{\text{sgn}\, t} = -i\,\text{pv } 1/\omega$, and so

$$\widehat{H(t)} = \pi\delta - i\,\text{pv } 1/\omega. \qquad (1.82)$$

At this point we show that the distributional Fourier transform maps L_2 into itself. For if $u \in L_2$ and $\varphi \in S$ then, by Schwarz's inequality and Parseval's relation on S, one has $|\langle \hat{u}, \varphi \rangle| = |\langle u, \hat{\varphi} \rangle| \leq \|u\|_{0,2}\|\hat{\varphi}\|_{0,2} = \text{const} \|\varphi\|_{0,2}$. Hence $\hat{u}$ is continuous on a subspace of L_2 and so may be extended to all of L_2. By the representation theorem of Riesz, quoted earlier, there exists a unique $v \in L_2$ such that $\langle \hat{u}, \varphi \rangle = \int_{-\infty}^{\infty} v\varphi \, dt = \int_{-\infty}^{\infty} u\hat{\varphi}$. Hence $v = \hat{u} \in L_2$. We use this fact to prove the following lemma.

Lemma 1.8. The Fourier transform is a one to one mapping of $\mathscr{D}'_{L_2}$ into the space S'_0 of tempered functions in L_2^{loc}.

PROOF. Let $u \in \mathscr{D}'_{L_2}$. Then $u = \Sigma_{j \leq l} D^j u_j$, $u_j \in L_2$, and $\hat{u} = \Sigma_{j \leq l} (i\omega)^j \hat{u}_j$. But $\hat{u}_j$ belongs to L_2 for $j \leq l$; thus $\hat{u}$ is contained in S'_0. Conversely, if $\hat{v} \in L_2^{loc}$, then $\mathfrak{F}^{-1}(i\omega)^l \hat{v} = D^l u$ for some $u \in L_2$, which proves the assertion.

The lemma just proven leads us to establish the next result which will be needed later.

Theorem 1.35. Every $u \in S'$ can be written as $(i\omega)^l v$ for some $l \geq 0$, where $v \in \mathscr{D}'_{L_2}$.

PROOF. Let $u = \mathfrak{F}(D^l(1 + t^2)^{m/2} u_0)$, according to Theorem 1.28; here $u_0 \in L_2$. Then u is, by (1.75), equal to $(i\omega)^l \mathfrak{F}((1 + t^2)^{m/2} u_0)$. But $(1 + t^2)^{m/2} u_0 \in S'_0$ and, by Lemma 1.8, its Fourier transform is in $\mathscr{D}'_{L_2}$, which is what we wished to prove.

Now an important result concerning the relation between Fourier transforms and convolution will be established.

Theorem 1.36. Let $u, v \in \mathscr{D}'_{L_2}$. Then $u * v$ exists and

$$\widehat{u * v} = \hat{u}\hat{v} \tag{1.83}$$

or equivalently if $u, v \in S'_0$ then $\widehat{uv} = (1/2\pi)\hat{u} * \hat{v}$.

PROOF. u, v can be written as $u = \Sigma_{j \leq l} D^l u_l$, $v = \Sigma_{i \leq k} D^i v_i$ and so, since $u_l * v_i$ certainly exists classically for $u_l, v_i \in L_2$, $D^{i+j}(u_l * v_i)$ is a well-defined distribution for each i, j, as is their sum which is $u * v$. Also one has classically (see Titchmarsh [T1]) $\widehat{u_l * v_i} = \hat{u}_l \hat{v}_i$. Hence $\mathfrak{F}D^{i+j}(u_l * v_i) = (i\omega)^{i+j} \hat{u}_l \hat{v}_i = (i\omega)^j \hat{u}_l (i\omega)^i \hat{v}_i$ so their sum is precisely $\hat{u}\hat{v}$.

The same type of argument also establishes the fact that

$$u * \text{pv} \, 1/\omega \in \mathscr{D}'_{L_2}$$

for any $u \in \mathscr{D}'_{L_2}$ for, again $u = \Sigma_{j \le l} D^j u_j$ and, since $u_j * \text{pv} \, 1/\omega = \text{pv} \int_{-\infty}^{\infty} [u_j(\zeta)/(\omega - \zeta)] \, d\zeta \in L_2$ (see Titchmarsh [T1], for this theorem of Riesz) then $u * \text{pv} \, 1/\omega$ is a sum of derivatives of L_2 functions, and so is in $\mathscr{D}'_{L_2}$.

Now note that since $\text{sgn} \, t \in S'_0$ then, using (1.81) and Theorem 1.35, $\text{pv} \, 1/\omega \in \mathscr{D}'_{L_2}$. In fact, since $\text{sgn} \, t/(1 + it) \in L_2$ then $\text{pv} \, 1/\omega = Dv_1 + v_2$, where v_1 and $v_2 \in L_2$. Hence $\text{pv} \, 1/\omega \in H'_{1,2}$.

To complete this section another useful result is proven.

Theorem 1.37. Let $u \in S' \cap \mathscr{D}'_+$. Then $u = D^l(1 + t^2)^{m/2} u_0$ for some l, m, and with a unique $u_0 \in L_2(0, \infty)$.

PROOF. We know from Theorem 1.28 that $u = D^l g$ for some $g \in S'_0$. Now let $g = g^+ + g^-$; then, by hypothesis, $D^l g = 0$ for $t < 0$, and so $D^l g^- = 0$ except, possibly, at the origin. Thus $D^l g^- = \Sigma_{j \le l-1} D^j \delta$ by virtue of Theorem 1.13 since g^- has at most a simple discontinuity at the origin. A particular solution of this equation is

$$H(-t) \sum_{j \le l-1} a_j t^j. \tag{1.84}$$

Since the complementary solution to the same equation is a polynomial of order $\le l - 1$ (Corollary 1.2) then g^- must have the form (1.84) since it vanishes for $t > 0$. Now let $f = g - \Sigma_{j \le l-1} a_j t^j$. Then $D^l f = D^l g$ and, moreover, $f \in \mathscr{D}'_+ \cap S'_0$. Thus f can be written as $(1 + t^2)^{m/2} u_0$ for some integer m, where $u_0 \in L_2(0, \infty)$. Note that the $u_0 \in L_2(0, \infty)$ is uniquely determined by l and m, since, if there were two, subtraction would yield $D^l[(1 + t^2)^{m/2}(u_{0_1} - u_{0_2})] = 0$, or, by Corollary 1.2, $u_{0_1} - u_{0_2}$ would be a polynomial which must be zero since u_{0_1} and u_{0_2} were assumed to be zero for $t < 0$.

CHAPTER II

The Laplace Transform

2.1 LAPLACE TRANSFORMS OF DISTRIBUTIONS WITH ARBITRARY SUPPORT

In this chapter we will establish the basic results concerning distributional Laplace transforms. The classical Laplace transform is defined for certain restricted classes of functions and the inversion is, similarly, limited. In the distributional setting, however, the Laplace transform achieves a certain structural completeness that ensures a greater flexibility in applications. In the next chapter the full scope of the distributional Laplace transform will be used in an essential way.

Most of the basic ideas of this chapter are due to Schwartz [Sc2] although there are many differences in detail in the present version. On occasion it will be necessary to refer to certain standard results concerning the classical Laplace transforms; for this purpose we refer to Widder's work [W1].

It is appropriate to begin with a definition of what we mean by Laplace transform for distributions. First, however, let us remark that from now on all distributions will have their support on the real t axis while their transforms are defined on the real ω axis. If there is any chance for confusion we write f_ω, for example, to denote dependence on ω.

DEFINITION 2.1. Let $f \in \mathcal{D}'$. Then, whenever $e^{-\sigma t}f \in S'$, the Laplace transform of f is defined by $\mathfrak{F}(fe^{-\sigma t})$ and we write $\mathcal{L}f$; σ is a real number.

Note that the set Γ of σ values for which $e^{-\sigma t}f \in S'$ may be empty. When it is not empty one has the following lemma.

Lemma 2.1. Γ is a nonzero interval on the real σ axis.

PROOF.   Let $\sigma_1, \sigma_2 \in \Gamma$ and suppose $\sigma = \lambda \sigma_1 + (1 - \lambda)\sigma_2$, $0 \le \lambda \le 1$; then σ lies on the line joining σ_1, σ_2. We must show that $\sigma \in \Gamma$. Now $e^{-\sigma t} = (e^{-\sigma_1 t})^\lambda (e^{-\sigma_2 t})^{1-\lambda}$. Suppose, without loss of generality, that $e^{-\sigma_1 t} < e^{-\sigma_2 t}$. Then $(e^{-\sigma_1 t})^\mu < (e^{-\sigma_2 t})^\mu$ for any $0 \le \mu \le 1$. Therefore, $e^{-\sigma_1 t} < (e^{-\sigma_1 t})^\lambda (e^{-\sigma_2 t})^{1-\lambda} < e^{-\sigma_2 t}$ so that, in general, $\alpha(\sigma, t) = e^{-\sigma t}/(e^{-\sigma_1 t} + e^{-\sigma_2 t})$ is a bounded and infinitely differentiable function in t, $0 < \alpha(\sigma, t) \le 1$. Similarly, all the derivatives of α with respect to t are given by expressions of the same type and so are themselves bounded; i.e., $\alpha \in \mathcal{D}_{L_\infty}$. Now the product of α with any $u \in S'$ is itself in S' (proof: $\langle \alpha u, \varphi \rangle = \langle u, \alpha \varphi \rangle$ for all $\varphi \in S$; but $\alpha \varphi \in S$ and if $\varphi_n \to 0$ in S then $\alpha \varphi_n \to 0$ so that $\alpha u \in S'$). Hence $e^{-\sigma t}f = \alpha(\sigma, t)(e^{-\sigma_1 t} + e^{-\sigma_2 t})f \in S'$. This proves the lemma.

Whenever f is a function, $\mathcal{L}f$ becomes the integral $\langle f e^{-\sigma t}, e^{-i\omega t} \rangle = \langle f, e^{-(\sigma + i\omega)t} \rangle$ which, of course, is the classical definition if the integral has meaning. In what follows we let p denote the complex variable $\sigma + i\omega$. Then one has the following lemma.

Lemma 2.2.   Let $f \in \mathcal{D}'_+$ and suppose Γ is the semi-infinite interval defined by $\sigma > \sigma_0$. If $fe^{-\sigma t} \in S'$ for $\sigma > \sigma_0$ then

$$\mathcal{L}f = \langle f, e^{-pt} \rangle \equiv \hat{f}(p) \qquad \text{Re } p > \sigma_0. \qquad (2.1)$$

PROOF.   First of all we have to show that $\langle f, e^{-pt} \rangle$ has meaning. Let $\lambda(t) \in C^\infty$ be equal to one on the support of f and vanish to the left of the origin. If $\sigma_1 > \sigma_0$ then $e^{-\sigma_1 t}f \in S'$, while $\lambda(t)e^{-(p-\sigma_1)t} \in S$ for Re $p > \sigma_1$ and so, independent of how λ was chosen, $\langle e^{-\sigma_1 t}f, \lambda(t)e^{-(p-\sigma_1)t} \rangle$ well defines $\langle f, e^{-pt} \rangle$ for Re $p > \sigma_1$. Now let $p' = p - \sigma_1$; then Re $p' > 0$ whenever Re $p > \sigma_1$. Also let $g = e^{-\sigma_1 t}f$ so that $g \in S'_+ = S' \cap D'_+$. Since S is dense in S' (Theorem 1.31) there exists a sequence $\{g_n\} \in S$ such that g_n, and hence λg_n, converge to g in the strong S' sense. Now $\lambda(t)\,e^{-p't} \in S$ for $\operatorname{Re} p' > 0$ and so $\langle \lambda g_n, e^{-p't}\rangle \to \langle g, e^{-p't}\rangle$ uniformly on every compact subset of the ω axis: $\{e^{-p't}\}$ is a bounded set in S as p' ranges over all compact subsets of the half plane $\operatorname{Re} p' > 0$. Hence for all $\varphi \in C_0^\infty$ one has $\langle \mathfrak{F}(\lambda g_n\, e^{-\sigma't}), \varphi\rangle = \langle\langle \lambda g_n, e^{-p't}\rangle, \varphi\rangle \to \langle\langle g, e^{-p't}\rangle, \varphi\rangle$ for $\sigma' = \operatorname{Re} p'$ [note that (2.1) is certainly valid for $f \in S$). But $\langle \mathfrak{F}(\lambda g_n\, e^{-\sigma't}), \varphi\rangle \to \langle \mathfrak{F}(g\, e^{-\sigma't}), \varphi\rangle$ as $n \to \infty$ since the Fourier transform is continuous on S'. Therefore $\hat f(p) \equiv \langle f, e^{-pt}\rangle = \langle f e^{-\sigma_1 t}, e^{-(p-\sigma_1)t}\rangle = \langle g, e^{-p't}\rangle = \mathfrak{F}(g\, e^{-\sigma't}) = \mathfrak{F}(f e^{-\sigma t})$ since $\sigma' = \sigma - \sigma_1$. But $\sigma_1 > \sigma_0$ is arbitrary and so $\langle f, e^{-pt}\rangle = \mathfrak{F}(f e^{-\sigma t})$ for $\operatorname{Re} p = \sigma > \sigma_0$. The next theorem is basic.

Theorem 2.1. Let Γ be an open interval of the real axis and $f \in \mathscr{D}'$ be such that $e^{-\sigma t} f \in S'$ for $\sigma \in \Gamma$. Then $\mathscr{L} f$ is a holomorphic function of $\sigma + i\omega$ in the infinite strip defined by $\sigma \in \Gamma$. Moreover, $\hat f(p) = \mathscr{L} f$ satisfies the boundedness condition

$$|f(\sigma + i\omega)| \le |\mathscr{P}_K(p)| \tag{2.2}$$

for some polynomial $\mathscr{P}_K$ which depends on the compact subset K as K varies over Γ.

Conversely, if $\hat f(p)$ is holomorphic in the strip defined by Γ, and if it satisfies the boundedness condition (2.2) on each compact subset K of Γ, then it is the Laplace transform of some $f \in \mathscr{D}'$ such that $f e^{-\sigma t} \in S'$ for $\sigma \in \Gamma$.

PROOF. Let $\sigma \in K \subset \Gamma$. Then $\langle \mathfrak{F}(f e^{-\sigma t}), \varphi\rangle$ can be written as $\langle \mathfrak{F}(f^+ e^{-\sigma t}), \varphi\rangle + \langle \mathfrak{F}(f^- e^{-\sigma t}), \varphi\rangle$, for $\varphi \in S$, by virtue of Corollary 1.4. But $f^+ e^{-\sigma t} = \lambda(t)(f^+ e^{-\sigma_1 t})e^{-(\sigma-\sigma_1)t}$, where $\lambda \in C^\infty$ equals one in $[0, \infty)$. Moreover, $\lambda e^{-(\sigma-\sigma_1)t} \in S$ for all $\sigma > \sigma_1$; hence $f^+ e^{-\sigma t} \in S'_+$ for all $\sigma > \sigma_1$ and, similarly, $f^- e^{-\sigma t} \in S'_-$ for all $\sigma < \sigma_2$. Therefore, by Lemma 2.2, $\hat f^+(p) \equiv \mathfrak{F}(f^+ e^{-\sigma t}) = \langle f^+, e^{-pt}\rangle$ for $\operatorname{Re} p > \sigma_1$, and we can also write it as $\langle (f^+ e^{-\sigma_1 t}), e^{-(p-\sigma_1)t}\rangle$. But $f^+ e^{-\sigma_1 t} \in S'_+$ and so, by Theorem 1.28, $f^+ e^{-\sigma_1 t} = D^l f_1^+$, where f_1^+ is a continuous function of polynomial growth with support in $[0, \infty)$. Thus $\langle f^+, e^{-pt}\rangle = \langle D^l f_1^+, e^{-(p-\sigma_1)t}\rangle = (p - \sigma_1)^l \langle f_1^+, e^{-(p-\sigma_1)t}\rangle$. By the classical theorem concerning Laplace transforms (cf. [W1]), the integral $\langle f_1^+, e^{-(p-\sigma_1)t}\rangle$

is holomorphic for Re $p > \sigma_1$, and uniformly bounded in every closed half plane Re $p \geq \sigma' > \sigma_1$; hence $\langle f^+, e^{-pt} \rangle$ is also holomorphic in that half plane and bounded uniformly by a polynomial (of order l) for Re $p \geq \sigma' > \sigma_1$. By the same reasoning $\langle f^-, e^{-pt} \rangle$ is holomorphic for $\sigma < \sigma_2$ and bounded by a polynomial in every closed half plane to the left of σ_2. Hence if $\hat{f}^-(p) \equiv \langle f^-, e^{-pt} \rangle$ then $\hat{f}(p) = \hat{f}^+(p) + \hat{f}^-(p)$ is holomorphic in the strip defined by the convex hull of K, and is uniformly bounded on every compact subset of K by a polynomial (note that integer l which defined the order of this polynomial depends on K). We can repeat this argument for a sequence of compact sets which exhaust Γ, and obtain a single function $\hat{f}(p)$, holomorphic in the strip defined by Γ, which satisfies the theorem [note that if $K \subset K_1$ then the $\hat{f}_1(p)$ defined by K_1 coincides with the $\hat{f}(p)$ given by K since they both equal $\mathfrak{F}(f e^{-\sigma t})$ for $\sigma \in K$].

Conversely, choose any inner strip defined by the convex hull of some compact $K \subset \Gamma$. There the order of $\mathscr{P}_K$ may be supposed to be l and so, since $\hat{f}(p)/p^{l+2} \in L_1$ as a function of ω for σ fixed in K, we can apply the classical inversion to obtain a continuous function

$$f_K(t) = \frac{1}{2\pi i} \int_{\sigma-i\infty}^{\sigma+i\infty} \frac{\hat{f}(p)e^{pt}}{p^{l+2}} \, dp \tag{2.3}$$

such that $e^{-\sigma t} f_K(t)$ is bounded (see Widder [W1] for details). Note that f_K does not vary with $\sigma \in K$ [briefly, form a rectangular contour with two of the parallel sides defined by fixed $\sigma, \sigma_1 \in K$. By Cauchy's theorem the sum of the integrals along opposite parallel sides are equal. As the rectangle is elongated to infinity the integrals vanish, which shows that (2.3) along σ_1 has the same value as the integral along σ]. Thus $\hat{f}(p)/p^{l+2}$ is the distributional Fourier transform of $e^{-\sigma t} f_K \in S'$ or $\hat{f}(p)/p^{l+2} = \mathfrak{F}(e^{-\sigma t} f_K) = \mathscr{L} f_K = \langle f_K, e^{-pt} \rangle$ for Re $p \in K$. However, $\langle D^{l+2} f_K, e^{-pt} \rangle = p^{l+2} \langle f_K, e^{-pt} \rangle = \hat{f}(p)$. Now exhaust Γ by an increasing sequence of compact sets K and repeat the argument on each such K. If $K' \supset K$ then $\hat{f}(p) = \mathfrak{F}(e^{-\sigma t} f_{K'}) = \mathfrak{F}(e^{-\sigma t} f_K)$ for all $\sigma \in K$ so that $\langle f_K - f_{K'}, e^{-\sigma t}\varphi \rangle = 0$ for all $\varphi \in C_0^\infty$ by virtue of the uniqueness of the Fourier transform. Since every $\psi \in C_0^\infty$ can be written as $e^{-\sigma t}\varphi$ for some other $\varphi \in C_0^\infty$ (proof: $\psi e^{\sigma t} \in C_0^\infty$) this implies

$f_K = f_{K'}$. Hence we have defined a single $f \in \mathcal{D}'$ such that $f e^{-\sigma t} \in S'$ and for which $\mathcal{L} f = \hat{f}(p)$, $\sigma \in \Gamma$.

As immediate corollaries of this proof one has the following corollary.

Corollary 2.1.   (Uniqueness of Laplace Transform.) If $\mathcal{L} f = \mathcal{L} g$ in Γ then $f = g$.

Corollary 2.2.   If $f e^{-\sigma t} \in S'$ for $\sigma \in \Gamma$ then $\mathfrak{F}(f e^{-\sigma t}) = \langle f, e^{-pt} \rangle$.

Corollary 2.3.   If $f e^{-\sigma t} \in S'$ for $\sigma \in \Gamma$ then $\mathcal{L} f = \hat{f}^+(p) + \hat{f}^-(p)$ where $\hat{f}^+, \hat{f}^-$ are analytic in overlapping half planes.

Corollary 2.4.   If $\hat{f}(p) = \mathcal{L} f$ then $p^l \hat{f}(p) = \mathcal{L}(D^l f)$.

The question now arises as to what happens as $\sigma \to \sigma_1$ or $\sigma \to \sigma_2$, where σ_1 and σ_2 are the end points of Γ. To answer this question we must extend the notion of Fourier transform to all of $\mathcal{D}'$. The essential ideas here are due to Gelfand and Shilov [G1].

First of all one observes that if $\hat{\varphi} \in C_0^\infty$ then its inverse Fourier transform can be extended as an entire function. In fact, since $e^{ipt} \hat{\varphi}(t)$ is analytic in p for each fixed t, then

$$\varphi(p) = \frac{1}{2\pi} \int_{-\infty}^{\infty} \hat{\varphi}(t) e^{ipt} \, dt \tag{2.4}$$

is an analytic function in the entire plane and yields $\varphi(\omega)$ when $\sigma = 0$. Integrating (2.4) by parts l times one then obtains

$$(ip)^l \varphi(p) = \frac{1}{2\pi} \int_{-\infty}^{\infty} D^l \hat{\varphi} \, e^{ipt} \, dt \tag{2.5}$$

and since supp $\hat{\varphi}$ is a compact set K then

$$|p^l \varphi(p)| \leq C_l \, e^{\alpha |\omega|} \tag{2.6}$$

where α depends on K. Conversely, if $\hat{\varphi}$ is the Fourier transform of the restriction to the ω axis of an entire function $\varphi(p)$ satisfying (2.6) for some α, l then it can be shown that $\hat{\varphi} \in C_0^\infty$ (see, for example, Zemanian [Z1]).

DEFINITION 2.2. Z will denote the collection of all inverse Fourier transforms of functions in C_0^∞.

A topology can be introduced on Z in such a way that convergence to zero of a sequence $\{\varphi_n\}$ in Z means that

$$|p^l \varphi_n(p)| \le C_l \, e^{\alpha|\omega|} \qquad (2.7)$$

holds for all l and where C_l and α do not depend on n, and

$$\sum_{j \le l} \max_K |D^j \varphi_n| \to 0 \qquad (2.8)$$

for each l. From (2.7) and (2.8) it is clear that if $\varphi_n \to \varphi$ in Z then $\varphi \in Z$. It is also clear that $Z \subset S$ since $C_0^\infty \subset S$. However, C_0^∞ is not contained in Z except for the trivial function because objects in C_0^∞ cannot be analytic functions.

The correspondence between C_0^∞ and Z is summarized in the next theorem which we leave without proof (cf. [Z1] again).

Theorem 2.2. The Fourier transform is a one–one bicontinuous mapping of Z onto C_0^∞ in the sense that if $\varphi \in Z$ then $\hat{\varphi} \in C_0^\infty$, $\mathfrak{F}^{-1}\hat{\varphi} \in Z$ and $\mathfrak{F}(\mathfrak{F}^{-1}\hat{\varphi}) = \varphi$. Moreover if $\varphi_n \to 0$ in Z then $\hat{\varphi}_n \to 0$ in $\mathscr{D}$, and conversely.

Let Z' denote the collection of all continuous linear functionals on Z. It may be shown quite easily that Z is dense in S and that the topology of S is stronger than that of Z. Hence the restriction of each $u \in S'$ to Z is continuous and so $u \in Z'$ (note that Z' is not contained in $\mathscr{D}'$ since C_0^∞ is not contained in Z). The standard weak topology is introduced on Z' and we say that $\{u_n\}$ in Z' converge to zero in the weak Z' topology whenever $\langle u_n, \varphi \rangle \to 0$ for all $\varphi \in Z$. Thus we see that $Z \subset S \subset S' \subset Z'$.

At this point we are in a position to define what we mean by the Fourier transform of an arbitrary distribution. Let $\hat{\varphi} \in C_0^\infty$; then $\langle f, \hat{\varphi} \rangle$ is well defined for all $f \in \mathscr{D}'$ and so $\mathfrak{F}f$ is defined as that object in Z' for which

$$\langle \mathfrak{F}f, \varphi \rangle = \langle f, \hat{\varphi} \rangle. \qquad (2.9)$$

By an abuse of notation, $\mathfrak{F}f \equiv \hat{f}$ will continue to denote Fourier transforms of distributions even when they are not in S'. $\varphi_n \to 0$ iff $\hat{\varphi}_n \to 0$ in $\mathscr{D}$ so that $\langle f, \hat{\varphi}_n \rangle$ and $\langle \mathfrak{F}f, \varphi_n \rangle$ both tend to zero also and the correspondence between $\mathscr{D}'$ and Z' is continuous. In fact one has the following theorem.

Theorem 2.3. The Fourier transform is one–one bicontinuous mapping of $\mathscr{D}'$ onto Z'.

The relation (2.9) extends the Parseval relation (1.71). If $f \in S'$ then (2.9) still holds, of course, but now the transform so defined coincides with the Schwartz transform defined by (1.71).

Now suppose $f e^{-\sigma t} \in S'$ for $\sigma \in \Gamma$. Then, by Corollary 2.3, $\mathscr{L}f = \hat{f}^+(p) + \hat{f}^-(p)$ where $\hat{f}^+$ is analytic for $\operatorname{Re} p > \sigma_1$ and $\hat{f}^-$ analytic for $\operatorname{Re} p < \sigma_2, \sigma_1 < \sigma_2$. As $\sigma \to \sigma_1$ then $\hat{f}^-(\sigma + i\omega) \to \hat{f}^-(\sigma_1 + i\omega)$, uniformly on the line defined by σ_1 and hence in the Z' topology, trivially. Also $\langle \hat{f}^+(\sigma + i\omega), \varphi(\omega) \rangle = \langle \mathfrak{F}(f^+ e^{-\sigma t}), \varphi \rangle = \langle f^+, e^{-\sigma t} \hat{\varphi} \rangle \to \langle f^+, e^{-\sigma_1 t} \hat{\varphi} \rangle = \langle \mathfrak{F}(f^+ e^{-\sigma_1 t}), \varphi \rangle$ for all $\varphi \in Z$ since $e^{-\sigma t} \hat{\varphi} \to e^{-\sigma_1 t} \hat{\varphi}$ in the topology of $\mathscr{D}$. Hence $\hat{f}(\sigma + i\omega) \to \mathfrak{F}(f e^{-\sigma_1 t})$ in the Z' topology, and similarly as $\sigma \to \sigma_2$. We state this as Theorem 2.4.

Theorem 2.4. Let $f e^{-\sigma t} \in S'$ for $\sigma \in \Gamma$ with endpoints σ_1, σ_2. Then $\hat{f}(p) = \mathscr{L}f \to \mathfrak{F}(f e^{-\sigma_1 t})$, as $\sigma \to \sigma_1$, in the weak Z' topology. Also, in the same sense, $\hat{f}(p) \to \mathfrak{F}(f e^{-\sigma_2 t})$ as $\sigma \to \sigma_2$.

Before closing this section let us note that if $f e^{-\sigma t} \in S'$ for all $\sigma \in \bar{\Gamma}$ (the closure of Γ) then an examination of the proof of Theorem 2.1 will reveal that a single polynomial $\mathscr{P}$ will suffice to bound $\mathscr{L}f$, i.e., one which does not depend on the compact $K \subset \Gamma$ that is chosen. This is so since $f e^{-\sigma t}$ is in S' even when σ is an end point of Γ.

2.2   LAPLACE TRANSFORMS OF DISTRIBUTIONS IN $\mathscr{D}'_+$

Theorem 2.5. Let $f \in \mathscr{D}'_+$ and suppose $f e^{-\sigma t} \in S'$ for $\sigma > 0$. Then $\mathscr{L}f = \hat{f}(p)$ is analytic for $\operatorname{Re} p > 0$ and the boundedness condition (2.2) is satisfied for every compact subset K of the positive axis $\sigma > 0$. Conversely, if $\hat{f}(p)$ is analytic in that half plane and satisfies such a boundedness condition then it is the Laplace transform of some $f \in \mathcal{D}'_+$ for which $f e^{-\sigma t} \in S'$, $\sigma > 0$.

PROOF. The proof mimics that of Theorem 2.1 in all details but one. In proving the converse, an estimate on the inversion formula (2.3) shows that

$$|f_K(t)| \leq \text{const } e^{\sigma t}, \tag{2.10}$$

so that if $t < 0$ we let $\sigma \to +\infty$ and obtain that f_K is null for $t < 0$. Therefore $D^{l+2} f_K \in \mathcal{D}'_+$.

The same proof as given in Theorem 2.4 establishes

Theorem 2.6. Let $\hat{f}(p)$ be the Laplace transform of $f \in \mathcal{D}'_+$ for which $f e^{-\sigma t} \in S'$ when $\sigma > 0$. Then $\hat{f}(\sigma + i\omega) \to \mathfrak{F}f$ in the Z' topology as $\sigma \to 0$ [here, of course, $\mathfrak{F}f$ is interpreted according to (2.9)].

One should note, again, that $\mathscr{P}_K$ in the proof of Theorem 2.5 depends on K unless $f e^{-\sigma t} \in S'$ for $\sigma \geq 0$ (or, equivalently, if $f \in S' \cap \mathcal{D}'_+ \equiv S'_+$).

We just said that if $f \in S'_+$ then its Laplace transform is bounded in each half plane $\text{Re } p \geq \sigma' > 0$ by a single polynomial, i.e.,

$$|\hat{f}(\sigma + i\omega)| \leq A(\sigma)|p|^l \tag{2.11}$$

for some $l \geq 0$, and where $A(\sigma)$ depends on $\sigma > 0$ (a proof of this is given below). We call the class of all such half plane holomorphic functions by the name of H^+. The H^+ functions are functions whose S' boundary values will be studied extensively in Chapter 3 and include the Hardy H^2 functions. We come now to an important result which will be fundamental in our researches. The theorem in question is apparently due to Gärding (unpublished), according to Streater–Wightman [St1]. A different proof appeared, independently, in Beltrami–Wohlers [B2] although a variant, valid for Z', was earlier given by Lauwerier [L1]. In the classical L_2 setting, the theorem is due to Paley and Wiener [P1].

Theorem 2.7. If $f \in S'_+$ then its Laplace transform $\hat{f}(p) \in H^+$ and $\hat{f}(\sigma + i\omega) \to \hat{f}$ in the S' topology as $\sigma \to 0$. Conversely, if $f(p) \in H^+$

and if $\hat{f}(p)$ has an S' limit g as $\sigma \to 0$, then $f(p) = \mathscr{L}f$, $f \in S'_+$, and $g = \hat{f}$.

PROOF. An examination of the proof of Theorem 2.1 shows that if $f \in S'_+$ then $\mathscr{L}f$ is bounded by a single polynomial, uniformly on every closed half plane of Re $p > 0$. Hence $\hat{f}(p) \in H^+$. Now let $\varphi \in S$. Then $\langle \hat{f}(\sigma + i\omega), \varphi(\omega) \rangle = \langle e^{-\sigma t}f, \hat{\varphi} \rangle \to \langle f, \hat{\varphi} \rangle = \langle \hat{f}, \varphi \rangle$ since $\lambda(t) e^{-\sigma t}\hat{\varphi} \to \lambda(t)\hat{\varphi}$ (see Lemma 2.2) in the topology of S as $\sigma \to 0$. Conversely, we already know that if $\hat{f}(p) \in H^+$ then it is the Laplace transform of some $f \in \mathscr{D}'_+$. Moreover, since $\langle \hat{f}(\sigma + i\omega), \varphi(\omega) \rangle \to \langle g_\omega, \varphi \rangle$ for all φ, then, by the closure of S' in the weak topology, $g_\omega \in S'$. Therefore $g_\omega = \hat{h}$ for some $h \in S'$. Since the topology of Z is weaker than that of S then, by Theorem 2.6, $g_\omega = \hat{f}$. Hence $\hat{f} = \hat{h}$ and so $f \in \mathscr{D}'_+ \cap S'$, which completes the proof.

The theorem just proven has its counterpart for Laplace transforms in a strip whenever $f e^{-\sigma t} \in S'$ for $\sigma \in \overline{\Gamma}$. We omit the straightforward details.

Note that not every $\hat{f}(p) \in H^+$ has an S' boundary value, even though the Z' boundary value always exists. For example, consider

$$\hat{f}(p) = \int_0^\infty I_0(2\sqrt{t}) e^{-pt} dt = e^{1/p}/p, \tag{2.12}$$

where f is the Bessel function $I_0(2\sqrt{t})$. $\hat{f}(p)$ is certainly in H^+ but since $I_0(2\sqrt{t}) \sim \exp\sqrt{t}$ as $t \to \infty$ then $f \notin S'$. Hence, by virtue of Theorem 2.7, $e^{1/p}/p$ cannot have an S' limit as $\sigma \to 0$. Thus it is possible for an H^+ function to be the Laplace transform of something in $\mathscr{D}'_+$ which is not in S' (note that $H(t) \exp\sqrt{t} e^{-\sigma t} \in S'_+$ for all σ strictly greater than zero).

Finally, we want to given an example of an H^+ function that does have an S' boundary value. Let $\hat{f}(p) = \mathscr{L}H = 1/p$. Then $1/p \to \pi\delta - i\,pv\,1/\omega$ as $\sigma \to 0$ in S', using the relation (1.82) and Theorem 2.7.

2.3 LAPLACE TRANSFORMS OF DISTRIBUTIONS IN $\mathscr{E}'$

As an analog of (2.6) we prove here

Theorem 2.8. Let $f \in \mathscr{E}'$ with support K. Then $\mathscr{L}f$ is an entire function having the exponential growth

$$|\hat{f}(p)| \leq \text{const}\,(1 + |p|)^l\, e^{A|\sigma|}, \tag{2.13}$$

where $l \geq 0$ and A is a constant which depends on K.

PROOF. $\hat{f}(p) = \langle f, e^{-pt} \rangle$. Let $\lambda \in C_0^\infty$ be equal to one on supp f. Then, since $\lambda[e^{-(p+h)t} - e^{-pt}]/h$ has a limit in the $\mathscr{D}$ topology as $h \to 0$, the difference quotient $[\hat{f}(p + h) - \hat{f}(p)]/h \to D\hat{f}(p)$ as $h \to 0$, for all p; hence $\hat{f}(p)$ is holomorphic in the entire plane. Note that we have used here the fact that $\langle f, e^{-pt} \rangle = \langle \lambda f, e^{-pt} \rangle$, independent of the λ so chosen. Now, by Theorem 1.12, since $e^{-pt} \in \mathscr{E}$,

$$|\langle f, e^{-pt} \rangle| \leq C \sum_{j \leq l} \max_K |D^j\, e^{-pt}| \tag{2.14}$$

for some $l \geq 0$. But the right-hand side of (2.14) is less than or equal to constant

$$e^{A|\sigma|} \sum_{j \leq l} |p|^j \leq \text{const}\,(1 + |p|)^l\, e^{A|\sigma|};$$

A clearly depends on the magnitude of K.

The converse to Theorem 2.8 is also true. Since it will not be needed in this book it suffices to state the result here. The proof in the distributional setting is due to Schwartz and extends another L_2 result of Paley and Wiener [P1]. Our version is somewhat weaker than Schwartz theorem.

Theorem 2.9. Let $\hat{f}(p)$ be an entire function that satisfies the estimate (2.13) for some A, l. Then $\hat{f}(p)$ is the Laplace transform of some $f \in \mathscr{E}'$.

CHAPTER III

*Distributional Boundary Values of Analytic Functions**

Introduction

The motivation for this chapter is the general interest of applied mathematicians in the boundary behavior of analytic functions. The basic question of the relationship between the boundary values and the holomorphic functions which define them was considered in Chapter II when we studied the bond that exists between Fourier and Laplace transforms. One of the aims of this chapter is to examine the implication of this bond in order to explore more fully the general problem of boundary behavior.

In the development of the theory of linear operators, the concept of Green's function (operator kernel, weighting function, impulse response, etc.) has played a central role since it constitutes a unique and concise characterization of the operator. When the operator is also translation invariant, the representation simplifies to a convolution statement, and in this case one can obtain alternate repre-

---

* In this chapter, the superscript numbers refer to the Notes and Remarks to be found in Section 3.6 (e.g., Theorem 3.3).

sentations by employing Laplace or Fourier transforms. A study of the linear, translation invariant operator by means of its transformed Green's function (transfer function, etc.) will involve functions holomorphic in a half plane whenever the operator is such that its Green's function is contained in $\mathscr{D}'_+$. The physicist deals extensively with such operators and generally refers to them as causal. The holomorphic nature of the Laplace transforms imply that their behavior in any finite region or, for that matter, along any arc in the domain of analyticity uniquely determines their behavior in the entire domain. One questions whether an equally intimate bond exists between the Laplace transforms and their boundary values. Classically, this problem led to considering representations of the holomorphic functions in terms of their boundary values as, for example, in the use of a Cauchy integral. However, unless restrictive assumptions were made on the boundary values, such as Hölder continuity, severe difficulties developed within the classical framework. It is only when distributions are allowed as boundary "values" that many of the difficulties can be overcome, and a precise bond between these values and the holomorphic functions is now possible.

In this chapter we develop distributional extensions of the Cauchy and Poisson integral formulas, and apply them to a study of the representation, uniqueness, and analytic continuation of holomorphic functions in a half plane. In addition, distributional boundary value theorems which extend the reciprocal Hilbert transform statements of the classic L_2 theory are obtained and then applied to the study of passive or dissipative operators.

At the end of this chapter some bibliographic remarks are made which, hopefully, will enable the reader to pursue these questions further.

3.1 CAUSAL OPERATORS

A linear, continuous, translation invariant operator T which maps $u \in \mathscr{E}'$ into $v \in \mathscr{D}'$ may be represented by the convolution (Theorem 1.18)

$$v = h * u, \tag{3.1}$$

where the distribution h is the kernel of the operator, i.e.,

$$T\delta = h * \delta = h. \tag{3.2}$$

By analogy with certain classical situations, we choose to call the distribution h the generalized Green's function. If $h \in S'$, this representation can be mapped, via the Fourier transform, into the product statement

$$\hat{v} = \hat{h}\hat{u}, \tag{3.3}$$

which is well defined since $\hat{u} = \mathfrak{F}u$ is infinitely differentiable (Theorem 2.8), and $\hat{h} = \mathfrak{F}h \in S'$. If $\hat{h}$ is not contained in S' but $h\,e^{-\sigma t} \in S'$ for $\sigma_1 < \sigma < \sigma_2$ then the Laplace transform yields

$$\hat{v}(p) = \hat{h}(p)\hat{u}(p), \tag{3.4}$$

which, from the discussion in Chapter II, is well defined pointwise for all p in the strip $\sigma_1 < \operatorname{Re} p < \sigma_2$. In addition, $\hat{h}(p)$ is holomorphic in this strip so that $\hat{h}(p) = \hat{v}(p)/\hat{u}(p)$ constitutes the algebraic representation of the operator sometimes loosely referred to by us as the resolvent. The properties of the operator, or physical system which it represents, are thus displayed in terms of a single holomorphic function $\hat{h}(p)$.

The physicist or engineer encounters many systems whose behavior is said to be causal or nonanticipatory. In the context of its operator representation, this property may be defined precisely in

DEFINITION 3.1. An operator T which maps $u \in \mathscr{E}'$ into $v \in \mathscr{D}'$ is said to be causal if for all $\varphi \in C_0^\infty$ whose support is contained in the semi-infinite interval $x < a$

$$\langle u_1, \varphi \rangle = \langle u_2, \varphi \rangle$$

implies that

$$\langle v_1, \varphi \rangle = \langle v_2, \varphi \rangle,$$

where u_1 and u_2 are arbitrary members of $\mathscr{E}'$.

It is clear that the operator represents a system where outputs do not anticipate the inputs. This property reflects itself in the Green's function in a very unique manner as demonstrated in the following

Theorem 3.1. A linear, continuous, translation invariant operator T which maps $u \in \mathscr{E}'$ into $v \in \mathscr{D}'$ is causal, in the sense of Definition 3.1, if and only if its Green's function $h = T\delta \in \mathscr{D}'_+$.

PROOF.   The necessity follows immediately on the basis of the convolution representation (1.1) since $u = 0$ implies $v = 0$, and $u = \delta$ yields $v = h$, whereas for all $\varphi \in C_0^\infty$ whose support is contained in $x < 0$

$$\langle 0, \varphi \rangle = \langle \delta, \varphi \rangle$$

so that

$$\langle 0, \varphi \rangle = \langle h, \varphi \rangle$$

for all such φ, i.e., by definition $h \in \mathscr{D}'_+$. The sufficiency may be established by considering

$$\langle v_1 - v_2, \varphi \rangle = \langle h * (u_1 - u_2), \varphi \rangle = \langle h_y, \langle (u_1 - u_2), \varphi(x + y) \rangle \rangle$$

and noting that if

$$\langle (u_1 - u_2), \varphi \rangle = 0$$

for all φ whose support is contained in $x < a$, the support of $\langle (u_1 - u_2), \varphi(x + y) \rangle$ is contained in $y < 0$, and

$$\langle h, \langle (u_1 - u_2), \varphi(x + y) \rangle \rangle = \langle (v_1 - v_2), \varphi \rangle = 0$$

for all such φ since $h \in \mathscr{D}'_+$.

Note that if an operator is causal then in particular it maps $\mathscr{D}'_+$ into itself. An immediate consequence of Theorem 3.1 is the fact that the algebraic representation (3.4) is defined, if it exists at all, in some half plane. That is, if $h\,e^{-\sigma t} \in S'$ for some σ_1, $h \in \mathscr{D}'_+$ implies that $h\,e^{-\sigma t} \in S'$ for all $\sigma \geq \sigma_1$ and the Laplace transform $\hat{h}$ is then defined and holomorphic in Re $p > \sigma_1$. If we assume that $\sigma_1 = 0$, we then encounter the situation in which the Laplace transform $\hat{h}(p)$ is holomorphic in the half plane Re $p > 0$, the Fourier transform $\hat{h}_\omega$ exists, and by Theorem 2.7, the Fourier transform is the boundary value of this holomorphic function. That is, the Laplace transform converges in the S' topology to a boundary value which is the Fourier transform. Thus in the study of tempered causal operators, we have available a characterization of the operator in terms of the distributional boundary values of a holomorphic function. It is this characterization which we wish to exploit in the remainder of this chapter.

3.2 Representation Theorems for Half-Plane Holomorphic Functions with S' Boundary Behavior

The role played by the Cauchy integral formula as a representation of holomorphic functions in terms of their boundary values is still central when the boundary behavior is distributional. We will obtain Cauchy formulas which are applicable to those half-plane functions that acquire boundary values in the S' topology. Not only will these formulas extend the range of applicability of the Cauchy representation, but they will be used as the basis of a discussion of certain uniqueness and analytic continuation questions to be pursued later in this chapter.

Initially, we will consider distributions f_ω that are contained in $\mathscr{D}'_{L_2}$ and, in an analogous fashion with the classic version, define the generalized Cauchy integral of f_ω by

$$C(f_\omega, p) = \frac{1}{2\pi} \left\langle f_\zeta, \frac{1}{p - i\zeta} \right\rangle. \tag{3.5}$$

The following theorem summarizes certain properties of this formula. First, however, we introduce the notation $f(p) \overset{S'}{\to} f_\omega$ to mean that $f(p)$ tends to f_ω in the S' topology as $\sigma \to 0$.

Theorem 3.2. Let $f_\omega \in \mathscr{D}'_{L_2}$. Then $C(f_\omega, p)$ exists, is holomorphic for $\operatorname{Re} p \neq 0$, and $C(f_\omega, p)$ converges in the S' topology (in either half plane) to distributions contained in $\mathscr{D}'_{L_2}$. If, in addition, f_ω is the S' boundary value of a function $f(p) \in H^+$ in $\operatorname{Re} p > 0$, then $C(f_\omega, p) = f(p)$ and so $C(f_\omega, p) \overset{S'}{\to} f_\omega$ in $\operatorname{Re} p > 0$. Similarly, if f_ω is the S' boundary value of a function $f(p) \in H^+$ in $\operatorname{Re} p < 0$, then $C(f_\omega, p) = -f(p)$ and so $-C(f_\omega, p) \overset{S'}{\to} f_\omega$ in $\operatorname{Re} p < 0$.

Proof. For every fixed p with $\operatorname{Re} p \neq 0$, $1/(p - i\zeta) \in \mathscr{D}_{L_2}$ so that $C(f_\omega, p) = 1/2\pi \langle f, 1/(p - i\zeta) \rangle$ exists when $f \in \mathscr{D}'_{L_2}$. Moreover, for

Re $p \neq 0$,

$$\left[\frac{1}{p' - i\zeta} - \frac{1}{p - i\zeta} \right] \frac{1}{p' - p}$$

tends to $-1/[(p - i\zeta)^2]$ in the $\mathscr{D}_{L_2}$ topology as $p' \to p$ (recall that $\varphi_n \to \varphi$ in the $\mathscr{D}_{L_2}$ topology if $\int_{-\infty}^{\infty} |D^k(\varphi_n - \varphi)| \, dx \to 0$ for all $k \geq 0$). Hence, by continuity,

$$\frac{d}{dp} \left\langle f, \frac{1}{p - i\zeta} \right\rangle = \lim_{p' \to p} \left\langle f, \left\{ \frac{1}{p' - i\zeta} - \frac{1}{p - i\zeta} \right\} \frac{1}{p' - p} \right\rangle$$

$$= \left\langle f, -\frac{1}{(p - i\zeta)^2} \right\rangle$$

so that $C(f, p)$ is holomorphic for Re $p \neq 0$. Now define $\mathfrak{F}^{-1}f_\omega = f_0$ and note that $f_0 \in S_0'$ since $f_\omega \in \mathscr{D}_{L_2}'$. From our definition of the Laplace transform we have $\mathscr{L}(f_0 H(t)) \equiv \mathfrak{F}(f_0 e^{-\sigma t}H)$ for $\sigma > 0$ where $H(t)$ is the Heaviside function. However, for $\sigma > 0$, $e^{-\sigma t}H \in S_0'$ and $\mathfrak{F}(e^{-\sigma t}H) = 1/(\sigma + i\omega)$ so that we may apply Theorem 1.36 to obtain

$$\mathscr{L}(f_0 H) \equiv \mathfrak{F}(f_0 e^{-\sigma t}H) = \frac{1}{2\pi} f_\omega * \frac{1}{\sigma + i\omega}$$

$$\equiv \frac{1}{2\pi} \left\langle f_\zeta, \frac{1}{p - i\zeta} \right\rangle \equiv C(f_\omega, p),$$

where $p = \sigma + i\omega$ and we restrict ourselves to Re $p > 0$. Note that the inner product representation of the convolution is valid since $1/(\sigma + i\omega) \in \mathscr{D}_{L_2}$ for $\sigma \neq 0$ and $f_\omega \in \mathscr{D}_{L_2}'$. Similarly we obtain

$$\mathscr{L}(f_0 H(-t)) \equiv \mathfrak{F}(f_0 e^{-\sigma t}H(-t)) = -\frac{1}{2\pi} f_\omega * \frac{1}{\sigma + i\omega}$$

$$\equiv -\frac{1}{2\pi} \left\langle f_\zeta, \frac{1}{p - i\zeta} \right\rangle \equiv -C(f_\omega, p)$$

for Re $p < 0$. Now we conclude the first portion of the proof of the theorem by noting that since $f_0 \in S'$, Theorem 2.7 implies that the two Laplace transforms noted above, and therefore the two Cauchy integrals acquire their boundary values (which are contained in $\mathscr{D}'_{L_2}$ since $f_0 H(t)$ and $f_0 H(-t) \in S'_0$), in the respective half planes, in the S' topology.

If in addition to being in $\mathscr{D}'_{L_2}$, f_ω is the S' boundary value of an $f(p) \in H^+$ in $\operatorname{Re} p > 0$ then by Theorem 2.7, $\mathfrak{F}^{-1} f_\omega = f_0 \in \mathscr{D}'_+$ and we have $f(p) \equiv \mathscr{L} f_0 = \mathscr{L}(f_0 H) = C(f_\omega, p)$ in $\operatorname{Re} p > 0$. The convergence of $C(f_\omega, p)$ was established above but in this case, since $f_0 \in \mathscr{D}'_+$, $\mathscr{L} f_0 \xrightarrow{S'} f_\omega = \mathfrak{F} f_0$ or $C(f_\omega, p) \xrightarrow{S'} f_\omega$ in $\operatorname{Re} p > 0$ similar arguments apply to the alternate situation in which $f_0 \in \mathscr{D}'_-$ or $f(p) \in H^+$ in $\operatorname{Re} p < 0$.

A simple but useful consequence of the theorem is contained in the next corollary.

Corollary 3.1. The necessary and sufficient condition that $\hat{f}_\omega \in \mathscr{D}'_{L_2}$ be the S' boundary value from $\operatorname{Re} p > 0$ of a function $\hat{f}(p) \in H^+$, i.e., that $\hat{f}_\omega$ and $\hat{f}(p)$ be the Fourier and Laplace transforms, respectively, of a distribution contained in $\mathscr{D}'_+$, is that

$$C(\hat{f}_\omega, p) = 0, \qquad \operatorname{Re} p < 0, \qquad (3.6)$$

in which case

$$\hat{f}(p) = C(\hat{f}_\omega, p), \qquad \operatorname{Re} p > 0.$$

PROOF. As in the proof of Theorem 3.2 we obtain with $\hat{f}_\omega = \mathfrak{F} f$ that $\hat{f}(p) = \mathscr{L} f = C(\hat{f}_\omega, p)$ for $\operatorname{Re} p > 0$ and $\langle f, e^{-pt} H(-t) \rangle = -C(\hat{f}_\omega, p)$, for $\operatorname{Re} p < 0$. But by our assumption, f is a function vanishing for $t < 0$ so that $\langle f, e^{-pt} H(-t) \rangle = 0$ $\operatorname{Re} p < 0$, and the necessity of the theorem is established. The sufficiency follows by reversing the steps of the above proof, i.e., for $\hat{f}_\omega \in \mathscr{D}'_{L_2}$, $0 = -C(\hat{f}_\omega, p) = \langle f, e^{-pt} H(-t) \rangle$ in $\operatorname{Re} p < 0$ implies, in view of Theorem 2.7, that $f \in \mathscr{D}'_+$.

Another corollary of Theorem 3.2 concerns the differentiability of the Cauchy integrals and is an extension of the classic derivative formulas.

Corollary 3.2. Let $u \in \mathscr{D}'_{L_2}$. Then for $\operatorname{Re} p \neq 0$,

$$\frac{d^k}{dp^k} C(u, p) = \frac{(-1)^k k!}{2\pi} \left\langle u, \frac{1}{(p - i\zeta)^{k+1}} \right\rangle = (-i)^k C(D^k u, p). \quad (3.7)$$

Proof. If $u \in \mathcal{D}'_{L_2}$, $D^k u \in \mathcal{D}'_{L_2}$ and Theorem 3.2 states that $C(u, p)$ and $C(D^k u, p)$ exist. Moreover, for Re $p \neq 0$

$$C(D^k u, p) = \frac{1}{2\pi} \left\langle D^k u, \frac{1}{p - i\zeta} \right\rangle = \frac{(-i)^k k!}{2\pi} \left\langle u, \frac{1}{(p - i\zeta)^{k+1}} \right\rangle$$

$$= \frac{(i)^k}{2\pi} \frac{d^k}{dp^k} \left\langle u, \frac{1}{p - i\zeta} \right\rangle = (i)^k \frac{d^k}{dp^k} C(u, p),$$

where the multiple differentiation with respect to p may be justified by repeated application of the arguments used in the proof of Theorem 3.2.

Although the first two terms of (3.7) are to be expected in view of the classic version of this result, the latter equality is decidedly a distributional statement since $D^k u$ is a weak derivative.

An illuminating example of the behavior of Cauchy integrals is obtained when $u \in \mathcal{E}'$. Specifically, these Cauchy integrals form a unique representation of the space $\mathcal{E}'$, in terms of holomorphic functions. This correspondence is summarized in the following theorem.

Theorem 3.3.[1] Let $u_\omega \in \mathcal{E}'$ with supp u equal K, and let $f(p) \equiv C(u, p)$. Then

(1) $f(p)$ is holomorphic in the entire plane except, at most, on the compact set K of the $p = i\omega$ axis,
(2) $f(p) = O(|p|^{-1})$ as $|p| \to \infty$, and
(3) $f(p)$ converges, in either half plane, to a distributional boundary value in the S' topology.

Conversely, given an $f(p)$ which satisfies conditions (1)–(3), then it is the Cauchy integral of some $u_\omega \in \mathcal{E}'$ whose support is contained in the compact set K.

Proof. Let $\lambda(\zeta) \in C_0^\infty$ equal one on the compact set K, i.e., $\lambda(\zeta)$ is zero outside some arbitrary compact set $K' \supset K$. In Chapter I, Lemma 1.3, we saw that such functions exist and, since K' is arbitrary, the distance between K' and K can be made arbitrarily small. Then since $u_\omega \in \mathscr{E}'$,

$$C(u, p) = \frac{1}{2\pi} \left\langle u, \frac{1}{p - i\zeta} \right\rangle = \frac{1}{2\pi} \left\langle u, \frac{\lambda(\zeta)}{p - i\zeta} \right\rangle.$$

But $d^n/dp^n[\lambda(\zeta)/(p - i\zeta)]$, as a function of ζ, is contained in $\mathscr{D}_{L_2}$ for all $n \geq 0$ and all $p \notin K$. A similar argument to that employed in Theorem 3.2 establishes the first condition. As to the second condition, $\lambda(\zeta)/(p - i\zeta)$ converges to zero in the topology of $\mathscr{E}$ as $|p| \to \infty$, so that $C(u, p)$ tends to zero for $|p| \to \infty$. However, $C(u, p)$ is holomorphic in the entire plane, with the exception of the compact set K of the $p = i\omega$ axis or, in particular, its Laurent expansion around the point at infinity exists in the form $C(u, p) = a_0 + (a_1/p) + \cdots$. The fact that $C(u, p)$ vanishes as $|p| \to \infty$ implies that $a_0 = 0$, and the order statement of condition (2) follows. The last condition is justified as a special case of Theorem 3.2.

Conversely, suppose that a given $f(p)$ satisfies the three conditions of the theorem. Then condition (3) implies that $f(p) \overset{S'}{\to} f_\omega^+$ in $\operatorname{Re} p > 0$, and $f(p) \overset{S'}{\to} f_\omega^-$ in $\operatorname{Re} p < 0$, so that we may define $u_\omega = f_\omega^+ - f_\omega^-$ and note that $C(u, p) = C(f^+, p) - C(f^-, p)$. However, f_ω^+ is the S' boundary value of $f(p)$ in $\operatorname{Re} p > 0$ so that Theorem 3.2 and Corollary 3.1 imply that

$$C(f^+, p) = \begin{array}{ll} f(p) & \operatorname{Re} p > 0 \\ 0 & \operatorname{Re} p < 0. \end{array}$$

Similarly,

$$-C(f^-, p) = \begin{array}{ll} f(p) & \operatorname{Re} p < 0 \\ 0 & \operatorname{Re} p > 0, \end{array}$$

or we have established that $f(p)$ is a Cauchy integral of some u, i.e., $C(u, p) = f(p)$, $\operatorname{Re} p \neq 0$. But $f(p)$ was assumed to be holomorphic for $p = i\omega$, $\omega \notin K$, from which we may conclude that

$$\langle f_\omega^+, \varphi \rangle = \lim_{\sigma \to 0^+} \langle f(p), \varphi \rangle = \lim_{\sigma \to 0^-} \langle f(p), \varphi \rangle = \langle f_\omega^-, \varphi \rangle$$

for all $\varphi \in C_0^\infty$ whose support is disjoint from K. Thus $\langle (f_\omega^+ - f_\omega^-), \varphi \rangle$

$= \langle u, \varphi \rangle = 0$ for all such φ, i.e., $u \in \mathcal{E}'$, and has support contained in the compact set K.

As we have noted in the previous development, a natural extension of the Cauchy integral formula may be made for those boundary values contained in $\mathcal{D}'_{L_2}$. The extension to other classes such as S' is not possible in a direct manner. However, one can obtain formulas of the Cauchy type. To do this we require the following definition which, although involving a slight abuse of notation, will be extremely convenient in what follows.

DEFINITION 3.2. If $u_\omega \in S'$, then we define $u/(i\omega)^k$ to be a solution, when it exists, of the distributional equation

$$(i\omega)^k \chi = u.$$

If in particular, $\mathfrak{F}^{-1}u_\omega \in S'_+$ then χ is that unique solution, when it exists, such that $\mathfrak{F}^{-1}\chi \in S'_0 \cap \mathcal{D}'_+$.

That there exists at least one k for which the above definition is meaningful may be established by noting that if $u_\omega \in S'$, $\mathfrak{F}^{-1}u_\omega = u \in S'$ and by the S' representation theorem (Theorem 1.28), there exists some k for which $u = D^k u_0$ where u_0 is a tempered locally L_2 integrable function (functions in S'_0). Thus $u_\omega = (i\omega)^k \mathfrak{F}u_0$, where $\mathfrak{F}u_0 \in \mathcal{D}'_{L_2}$ is a solution to the distributional equation stated in Definition 3.2. Note that on compact sets disjoint from the origin $\chi = [u/(i\omega)^k]$ is well defined as the product of $[1/(i\omega)^k]$ and $u \in S'$. It is of interest to remark here that any two tempered solutions to this problem differ at most by a distribution of point support. For if $(i\omega)^k(\chi_1 - \chi_2) = 0$ then $D^k(\hat{\chi}_1 - \hat{\chi}_2) = 0$, which implies (see Corollary 1.2) that $\hat{\chi}_1 - \hat{\chi}_2$ is a polynomial of order $\leq k - 1$, which, in turn, shows that $\chi_1 = \chi_2 + \Sigma_{j \leq k-1} D^j \delta$.

Now, in particular, if $\mathfrak{F}^{-1}u_\omega \in S'_+$, then in view of Theorem 1.37 we can find some k for which $\mathfrak{F}^{-1}u_\omega = D^k u_0$, $u_0 \in S'_0 \cap \mathcal{D}'_+$, where u_0 is unique. Thus the solution $\chi = \mathfrak{F}u_0$ is unique.

In view of the foregoing discussion, we may extend the Cauchy integral to S' in the following manner.

Theorem 3.4. If $\hat{f}_\omega \in S'$ then for some k, $C(\hat{f}_\omega/(i\omega)^k, p)$ is holomorphic for $\mathrm{Re}\, p \neq 0$, and converges in the S' topology to a distribution $\in S'$ (in either half plane). If, in addition, $\hat{f}_\omega$ is the S' boundary value of a function $\hat{f}(p) \in H^+$ for $\mathrm{Re}\, p > 0$, then

$$\hat{f}(p) = p^k C(\hat{f}_\omega/(i\omega)^k, p) \tag{3.8}$$

and

$$D^k \hat{f}(p) = \frac{1}{2\pi} \left\langle \hat{f}_\zeta, \frac{d^k}{dp^k} \frac{1}{p - i\zeta} \right\rangle$$

$$= \frac{(i)^k}{2\pi} \left\langle \hat{f}_\zeta, \frac{d^k}{d\zeta^k} \frac{1}{p - i\zeta} \right\rangle \tag{3.9}$$

in $\mathrm{Re}\, p > 0$. Similarly, if $\hat{f}(p) \in H^+$, $\mathrm{Re}\, p < 0$, then

$$\hat{f}(p) = -p^k C(\hat{f}_\omega/(i\omega)^k, p)$$

and

$$D^k \hat{f}(p) = -\frac{1}{2\pi} \left\langle \hat{f}_\zeta, \frac{d^k}{dp^k} \frac{1}{p - i\zeta} \right\rangle$$

in $\mathrm{Re}\, p < 0$. In addition, $p^k C(\hat{f}_\omega/(i\omega)^k, p) \overset{S'}{\to} \hat{f}_\omega$ in $\mathrm{Re}\, p > 0$, and similarly for $\mathrm{Re}\, p < 0$.

PROOF. In view of the discussion associated with Definition 3.2, there exists at least one k for which $\hat{f}_\omega/(i\omega)^k$ is a well-defined distribution contained in $\mathscr{D}'_{L_2}$, and $\hat{f}_\omega = (i\omega)^k(\hat{f}_\omega/(i\omega)^k) = \mathfrak{F}(D^k f_0)$. In particular, when $\hat{f}_\omega$ is the S' limit of an H^+ function, then $\mathscr{L}f = p^k \mathscr{L}f_0$ and we can appeal directly to Theorem 3.2 to establish, with the necessary addition of the factor p^k, all the conditions of the present theorem except for the derivative statement. However, if we define $\hat{f}_\omega/(i\omega)^k = \hat{u}_\omega \in \mathscr{D}'_{L_2}$ then $\hat{f}_\omega = (i\omega)^k \hat{u}_\omega$ and $\hat{f}(p) = p^k \hat{u}(p)$. A direct application of Theorem 3.2 then yields $\hat{u}(p) = C(\hat{u}_\omega, p)$. Thus $D^k \hat{f}(p) = D^k(p^k \hat{u}) = D^k[1/2\pi \langle \hat{u}_\zeta, p^k/(p - i\zeta) \rangle]$. A similar continuity argument to that used in Theorem 3.2 allows us to write, finally,

$$D^k \hat{f}(p) = \frac{1}{2\pi} \left\langle \hat{u}_\zeta, \frac{d^k}{dp^k} \frac{p^k}{p - i\zeta} \right\rangle = \frac{1}{2\pi} \left\langle \hat{u}_\zeta, \frac{(i\zeta)^k(-1)^k k!}{(p - i\zeta)^{k+1}} \right\rangle$$

$$= \frac{1}{2\pi} \left\langle \hat{f}_\zeta, \frac{d^k}{dp^k} \frac{1}{p - i\zeta} \right\rangle = \frac{(i)^k}{2\pi} \left\langle \hat{f}_\zeta, \frac{d^k}{d\zeta^k} \frac{1}{p - i\zeta} \right\rangle$$

since

$$\frac{d^k}{dp^k} \left[\frac{p^k}{p - i\zeta} \right] = \frac{d^k}{dp^k} \left[p^{k-1} + (i\zeta)p^{k-2} + \cdots + (i\zeta)^{k-1} + \frac{(i\zeta)^k}{p - i\zeta} \right]$$

$$= (i\zeta)^k \frac{d^k}{dp^k} \frac{1}{p - i\zeta}.$$

Thus we have extended the Cauchy integral to objects in S' by obtaining two Cauchy "type" representations for analytic functions in H^+. The above theorem relies heavily on the fact that one can place distributions contained in S' into a one-to-one correspondence with distributions in $\mathscr{D}'_{L_2}$ via some suitable factor $(i\omega)^k$. This allows one to focus attention on the $\mathscr{D}'_{L_2}$ class and obtain similar results for S' by proper insertion of the p^k factors. In this spirit we consider the following useful representation theorem for holomorphic functions.

Theorem 3.5. If $\hat{f}_\omega \in \mathscr{D}'_{L_2}$ is the S' boundary value of a $\hat{f}(p) \in H^+$ in Re $p > 0$ then

$$\hat{f}(p) = \frac{1}{\pi} \left\langle \operatorname{Re} \hat{f}, \frac{1}{p - i\zeta} \right\rangle = \frac{i}{\pi} \left\langle \operatorname{Im} \hat{f}, \frac{1}{p - i\zeta} \right\rangle \qquad (3.10)$$

PROOF. In view of the conditions of this theorem, we may apply Corollary 3.1 to obtain $1/2\pi \langle \hat{f}, 1/(p - i\zeta) \rangle = 0$ Re $p < 0$ or, alternatively, $1/2\pi \langle \hat{f}, 1/(-\bar{p} - i\zeta) \rangle = 0$ Re $p > 0$. Separating the distribution $\hat{f}$ into its real and imaginary components allows the latter expression to be written as $1/2\pi \langle \operatorname{Re}\hat{f}, 1/(-\bar{p} - i\zeta) \rangle = -i/2\pi \langle \operatorname{Im} \hat{f}, 1/(-\bar{p} - i\zeta) \rangle$ in Re $p > 0$ or, taking conjugates, $1/2\pi \langle \operatorname{Re} \hat{f}, 1/(p - i\zeta) \rangle = i/2\pi \langle \operatorname{Im} \hat{f}, 1/(p - i\zeta) \rangle$ in Re $p > 0$. The theorem follows directly from the Cauchy integral, written as

$$\hat{f}(p) = \frac{1}{2\pi} \left\langle (\text{Re} \, \hat{f} + i \, \text{Im} \, \hat{f}), \frac{1}{p - i\zeta} \right\rangle, \quad \text{Re} \, p > 0.$$

The representation of Theorem 3.5 is of independent interest in the study of holomorphic functions because it displays the fact that either the real or imaginary component of the boundary behavior uniquely characterize the original holomorphic function. The classic L_2 versions of this theorem have been applied to problems in potential theory as a means of obtaining both harmonic and holomorphic half-plane functions that have specified boundary values. In this context, Theorem 3.5 yields solutions to the so-called Dirichlet half-plane problem.

Another application of the generalized Cauchy integral concerns the solution of what, in classic boundary-value studies, is referred to as a modified Hilbert problem. Specifically, if a prescribed function is known to be holomorphic in some strip of the complex plane ($\sigma_1 < \text{Re} \, p < \sigma_2$), the problem is to find two functions that are separately holomorphic in overlapping half planes $\text{Re} \, p > \sigma_1$, and $\text{Re} \, p < \sigma_2$, respectively, such that their sum in the strip is the original function. This so-called Hilbert decomposition is utilized in the solution of Wiener-Hopf integral equations. Here the problem is to represent a given function as the product of two functions that are required to be holomorphic in their respective half planes: one takes the logarithm of the given function, assuming it is holomorphic in the strip, forms a Hilbert decomposition, and the original problem is then satisfied by inverting the logarithmic operation to obtain a product form. In the Hilbert problem, one must consider the situation in which the strip degenerates into a line. Here, however, the same type of decomposition may be obtained if one deals with the boundary values of the two half-plane functions defined by the line. We will first consider the basic problem of obtaining a Hilbert decomposition in a strip and, in view of the previous discussion involving the equivalence which can be made between S' and $\mathscr{D}'_{L_2}$, we restrict ourselves to the $\mathscr{D}'_{L_2}$ class.[2] The basic result is summarized in the following theorem.

Theorem 3.6.   Let $f(p)$ be holomorphic and uniformly bounded by a fixed polynomial in every compact subset of the strip $\Gamma: \sigma_1 < \operatorname{Re} p < \sigma_2$. In addition, let $\hat{f}(p)$ converge in the S' topology (from the interior of the strip) to the boundary values $f_1 \in \mathscr{D}'_{L_2}$ on $\operatorname{Re} p = \sigma_1$ and $f_2 \in \mathscr{D}'_{L_2}$ on $\operatorname{Re} p = \sigma_2$. Then for $\sigma_1 < \operatorname{Re} p < \sigma_2$

$$\hat{f}(p) = C(f_1, p) - C(f_2, p) \tag{3.11}$$

where $C(f_1, p)$ is holomorphic in $\operatorname{Re} p > \sigma_1$, and $C(f_2, p)$ is holomorphic in $\operatorname{Re} p < \sigma_2$.

PROOF.   In the discussion of the Laplace transform in Chapter II, it was established that, when $f e^{-\sigma t} \in S'$ for $\sigma_1 \le \sigma \le \sigma_2$, $\mathscr{L}f = \hat{f}(p) = \hat{f}^+(p) + \hat{f}^-(p)$, where $\hat{f}^+$ and $\hat{f}^-$ are holomorphic and uniformly bounded by a polynomial in the respective half planes $\operatorname{Re} p > \sigma_1$, $\operatorname{Re} p < \sigma_2$. The converse of this statement, which was also established, allows us to immediately conclude that a Hilbert decomposition is possible in terms of $\hat{f}^+(p)$ and $\hat{f}^-(p)$. Moreover, the discussion in Chapter II implies that $\hat{f}^+$ and $\hat{f}^-$ converge in the S' topology to their respective boundary values which in the present case are assumed to be contained in $\mathscr{D}'_{L_2}$. Thus in view of Theorem 3.1 (after a translation of the axis)

$$\hat{f}^+(p) = C(\hat{f}^+_\omega, p - \sigma_1), \qquad \operatorname{Re} p > \sigma_1$$

and

$$\hat{f}^-(p) = -C(\hat{f}^-_\omega, p - \sigma_2), \qquad \operatorname{Re} p < \sigma_2$$

where $\hat{f}^+_\omega$ and $\hat{f}^-_\omega$ are the respective boundary values of $\hat{f}^+(p)$ and $\hat{f}^-(p)$. But from Corollary 3.1 we infer, due to the holomorphic behavior of $\hat{f}^+$ and $\hat{f}^-$, that

$$C(\hat{f}^-(\sigma_1 + i\omega), p - \sigma_1) = 0, \qquad \operatorname{Re} p > \sigma_1$$

and

$$C(\hat{f}^+(\sigma_2 + i\omega), p - \sigma_2) = 0, \qquad \operatorname{Re} p < \sigma_2,$$

so that we may write

$$\hat{f}^+(p) = C([\hat{f}^+_\omega + \hat{f}^-(\sigma_1 + i\omega)], p - \sigma_1), \qquad \operatorname{Re} p > \sigma_1$$

and

$$\hat{f}^-(p) = -C([\hat{f}_\omega^- + \hat{f}^+(\sigma_2 + i\omega)], p - \sigma_2), \qquad \mathrm{Re}\, p < \sigma_2.$$

But the latter representations yield the stated result of the theorem since $\hat{f}_\omega^+ + \hat{f}^-(\sigma_1 + i\omega)$ is the S' boundary value of $\hat{f}(p)$ on $\mathrm{Re}\, p = \sigma_1$ and $\hat{f}_\omega^- + \hat{f}^+(\sigma_2 + i\omega)$ is the S' boundary value of $\hat{f}(p)$ on $\mathrm{Re}\, p = \sigma_2$.

Turning our attention now to the degenerate strip, we obtain the desired representation expressed in (we assume for simplicity that the strip coincides with $p = i\omega$) Theorem 3.7.

Theorem 3.7. Let $\hat{f}_\omega \in S'$. Then $\hat{f}_\omega = \hat{f}_\omega^+ + \hat{f}_\omega^-$ where $\hat{f}_\omega^+ \in S'$ is the S' boundary value, in $\mathrm{Re}\, p > 0$, of the function $\hat{f}^+(p) = p^k C(\hat{f}/(i\omega)^k, p)$, and $\hat{f}_\omega^- \in S'$ is the S' boundary value in $\mathrm{Re}\, p < 0$, of the function $\hat{f}^-(p) = -p^k C(\hat{f}/(i\omega)^k, p)$.

PROOF. Since $\mathfrak{F}^{-1}\hat{f}_\omega \equiv f \in S'$ we may represent it (Theorem 1.28) as $f = D^k(f_0 H(t)) + D^k(f_0 H(-t))$. Applying Theorem 3.4, we obtain $\hat{f}^+(p) \equiv \mathscr{L}[D^k(f_0 H)] = p^k C(\hat{f}_\omega/(i\omega)^k, p) \overset{S'}{\to} \hat{f}_\omega^+ \equiv \mathfrak{F}(f_0 H)$ and $\hat{f}^-(p) \equiv \mathscr{L}[D^k(f_0 H(-t))] = -p^k C(\hat{f}_\omega/(i\omega)^k, p) \overset{S'}{\to} \hat{f}_\omega^- \equiv \mathfrak{F}[f_0 H(-t)]$. The theorem follows directly because $\mathfrak{F}[D^k(f_0 H)] + \mathfrak{F}[D^k(f_0 H(-t))] = \hat{f}_\omega$.

The explicit representation for the $\hat{f}_\omega^+$ and $\hat{f}_\omega^-$ of the preceding theorem is available in the following corollary.

Corollary 3.3. Let $\hat{f}_\omega, \hat{f}_\omega^+,$ and $\hat{f}_\omega^-$ be as defined in Theorem 3.7.[3] Then

$$\hat{f}_\omega^+ = \frac{(i\omega)^k}{2\pi i}\left[\hat{f}_\omega/(i\omega)^k * \mathrm{pv}\,\frac{1}{\omega}\right] + \frac{\hat{f}_\omega}{2} \tag{3.12}$$

and

$$\hat{f}_\omega^- = -\frac{(i\omega)^k}{2\pi i}\left[\hat{f}_\omega/(i\omega)^k * \mathrm{pv}\,\frac{1}{\omega}\right] + \frac{\hat{f}_\omega}{2}. \tag{3.13}$$

PROOF. From the proof of Theorem 3.7 we have $\hat{f}_\omega^+ = \mathfrak{F}[D^k(f_0 H)]$, which may be rewritten using Theorem 1.36, and the fact that

$$\mathfrak{F}[H(t)] = \pi\delta + \mathrm{pv}\,\frac{1}{i\omega},$$

as

$$\hat{f}_\omega^+ = (i\omega)^k \mathfrak{F}[f_0 H] = \frac{(i\omega)^k}{2\pi}\left[\mathfrak{F}f_0 * \left(\pi\delta + \mathrm{pv}\,\frac{1}{i\omega}\right)\right]$$

$$= \frac{(i\omega)^k}{2\pi i}\left[\hat{f}_\omega/(i\omega)^k * \mathrm{pv}\,\frac{1}{\omega}\right] + \frac{(i\omega)^k}{2}\left[\hat{f}_\omega/(i\omega)^k\right]$$

because $\mathfrak{F}f_0 = \hat{f}_\omega/(i\omega)^k$. Similarly,

$$\hat{f}_\omega^- = (i\omega)^k \mathfrak{F}[f H(-t)]$$

$$= \frac{(i\omega)^k}{2\pi}\left[\mathfrak{F}f * \left(\pi\delta - \mathrm{pv}\,\frac{1}{i\omega}\right)\right].$$

Another representation of holomorphic functions that has been utilized in classic studies, is the Poisson integral formula. One defines this integral in terms of Cauchy integrals, in the following manner: if $u \in \mathscr{D}'_{L_2}$, its Poisson integral is defined, for $\mathrm{Re}\,p \neq 0$ as,

$$P(u, p) = C(u, p) - C(u, -\bar{p}) = \frac{\sigma}{\pi}\left\langle u, \frac{1}{\sigma^2 + (\omega - \zeta)^2}\right\rangle \qquad (3.14)$$

Since for $u \in \mathscr{D}'_{L_2}$ the two Cauchy integrals are well defined, the Poisson integral also has meaning. Although in general $P(u, p)$ is not holomorphic in $\mathrm{Re}\,p \neq 0$ because $C(u, -\bar{p})$ fails to be holomorphic, the Poisson integral has a unique property not shared with the Cauchy integral; its distributional boundary value for $p = i\omega$ equals u. Formally, we state this as Theorem 3.8.

Theorem 3.8. Let $u \in \mathscr{D}'_{L_2}$. Then $P(u, p)$ is a harmonic function that converges in the S' topology (in $\mathrm{Re}\,p > 0$) to u.

PROOF. From Theorem 3.7, $C(u, p) \overset{S'}{\to} u_\omega^+$ in $\mathrm{Re}\,p > 0$, and $-C(u, p) \overset{S'}{\to} u_\omega^-$ in $\mathrm{Re}\,p < 0$. But

$$-C(u, p)|_{\mathrm{Re}\,p<0} = -C(u, -\bar{p})|_{\mathrm{Re}\,p>0}$$

or in $\mathrm{Re}\,p > 0$, $-C(u, -\bar{p})|_{\mathrm{Re}\,p>0} \overset{S'}{\to} u_\omega^-$. Thus $C(u, p) - C(u, -\bar{p})$ $\overset{S'}{\to} u_\omega^+ + u_\omega^- = u_\omega$ in $\mathrm{Re}\,p > 0$. The fact that $P(u, p)$ is harmonic in $\mathrm{Re}\,p > 0$ may be established directly.

A further bond between the Poisson and Cauchy integrals is illustrated by those situations in which u is itself the S' boundary value of an H^+ function. The relationship is summarized in

Theorem 3.9. $\hat{f}_\omega \in \mathscr{D}'_{L_2}$ is the S' boundary value in $\operatorname{Re} p > 0$ of a function $\hat{f}(p) \in H^+$ iff for $\operatorname{Re} p > 0$

$$\hat{f}(p) = C(\hat{f}_\omega, p) = P(\hat{f}_\omega, p).$$

PROOF. From Corollary 3.1, we obtain, since $\hat{f}_\omega$ is the boundary value of an H^+ function, $C(\hat{f}_\omega, p)|_{\operatorname{Re} p < 0} = C(\hat{f}_\omega, -\bar{p})|_{\operatorname{Re} p > 0} = 0$ or $P(\hat{f}_\omega, p) = C(\hat{f}_\omega, p) = \hat{f}(p)$, $\operatorname{Re} p > 0$.

At this point it may be desirable to note that although many of the statements obtained in this section have classical counterparts, if one restricts the functions to be in L_2 and employs a normed convergence to relate the boundary behavior to the original holomorphic function, the results are generalizations which include these classic results as special cases. Thus the class of admissible objects having Cauchy integrals, for example, has been enlarged and this accomplished in a way that preserves the essential structure or properties of these integrals. One must, however, deal with a weaker convergence if he is to give meaning to these results. But, rather than a handicap, in many respects this is an advantage since it allows for the more erratic behavior of the boundary values which is obscured in the classic setting. A simple example to illustrate these conclusions is the function $1/p \in H^+$. Classically, $\lim_{\sigma \to 0} 1/p = 1/i\omega$ a.e. but the pv Cauchy integral of $1/i\omega$ yields $\frac{1}{2} p$ for $\operatorname{Re} p > 0$, an inconsistent result. In the setting of distribution theory, however, this conflict is resolved since $1/p$ converges in the S' topology to the distribution $[\pi\delta + \operatorname{pv}(1/i\omega)] \in \mathscr{D}'_{L_2}$ and its Cauchy integral does reproduce the function $1/p$. The utility of the Cauchy and Poisson representation in the study of holomorphic functions is greatly enhanced by the distributional generalization since the class of holomorphic functions has been enlarged to the same degree as the class of admissible boundary values (note that $H^2 \subset H^+$ while $L_2 \subset \mathscr{D}'_{L_2} \subset S'$).

3.3   BOUNDARY VALUE THEOREMS AND GENERALIZED HILBERT TRANSFORMS

The principal concern of the previous section was the representation of holomorphic functions in terms of their distributional boundary values. The aim of this section is to investigate the boundary values themselves. In particular, we will describe techniques that allow one to ascertain directly if a given distribution is the boundary behavior of a half-plane holomorphic function. The precursors of these techniques are the L_2 theorems of Hille and Tamarkin [H1] and Titchmarsh [T1] which conclude that $f(\omega) \in L_2$ is the boundary value, in the L_2 norm and pointwise a.e., of some $f(p)$ holomorphic in Re $p > 0$, if and only if

$$\operatorname{Re} f(\omega) = \frac{1}{\pi} \operatorname{pv} \int_{-\infty}^{\infty} \frac{\operatorname{Im} f(\zeta)}{\omega - \zeta} d\zeta.$$

The condition obtained in the theorem may also be phrased as: the real and imaginary components of the L_2 boundary value of holomorphic functions must be Hilbert transform pairs. Our first task will be to extend these considerations to include those boundary values which are attained in the S' topology.

Before stating the basic theorem of this section we will establish the following lemma that will facilitate the future discussion.

Lemma 3.1. If $u \in \mathscr{D}'_{L_2}$ then $D^k[\omega^k(u * \operatorname{pv}(1/\omega))] = (\omega^k u) * D^k \operatorname{pv}(1/\omega)$.

PROOF.   Using the Leibnitz rule we have, for all $\varphi \in C_0^\infty$,

$$\left\langle D^k\left[\omega^k\left(u * \operatorname{pv}\frac{1}{\omega}\right)\right], \varphi \right\rangle$$

$$= \sum_{j \le k} \binom{k}{j} \left\langle u * D^{k-j} \operatorname{pv}\frac{1}{\omega}, D^j(\omega^k)\varphi(\omega) \right\rangle$$

$$= \left\langle u, \operatorname{pf} \int_{-\infty}^{\infty} \sum_{j \le k} \binom{k}{j} D^j_\omega(\omega + \zeta)^k D^{k-j}_\omega\left(\frac{1}{\omega}\right) \varphi(\omega + \zeta) \, d\omega \right\rangle$$

because $D^n \operatorname{pv}\frac{1}{\omega} = D^n \operatorname{pf}\frac{1}{\omega} = \operatorname{pf} D^n\frac{1}{\omega}$ [Chapter I, Eq. (1.25)]. But

$$\sum_{j \leq k} \binom{k}{j} D_\omega^j(\omega + \zeta)^k D_\omega^{k-j}\left(\frac{1}{\omega}\right) = \left[D_\omega^k \ (\omega + \zeta)^k \frac{1}{\omega} \right] =$$

$$D_\omega^k \left[\frac{\zeta^k}{\omega} + \text{polynomial of order } k-1 \text{ in } \omega \right] = \zeta^k D^k \frac{1}{\omega}.$$

Hence the entire expression reduces to $\langle \zeta^k u, \text{pf} \int_{-\infty}^{\infty} \varphi(\omega + \zeta) D^k(1/\omega) \, d\omega \rangle$
$= \langle (\omega^k u) * D^k \text{pv}(1/\omega), \varphi(\omega) \rangle$.

We are now in a position to generalize the classic boundary value statement in Theorem 3.10.

Theorem 3.10. The necessary and sufficient condition that $\hat{u}_\omega \in S'$ be the boundary value in the S' topology of a function $\hat{u}(p) \in H^+$, i.e., that $\hat{u}_\omega$ and $\hat{u}(p)$ be the Fourier and Laplace transforms respectively of a distribution in $\mathscr{D}'_+$, is that either

$$D^k \hat{u} = \frac{1}{\pi i} \hat{u} * D^k \text{pv} \frac{1}{\omega} \qquad (3.15)$$

or

$$\hat{u} = \frac{(i\omega)^k}{\pi i} \left[\frac{\hat{u}}{(i\omega)^k} * \text{pv} \frac{1}{\omega} \right] \qquad (3.16)$$

where k is defined as any integer for which $\mathscr{F}^{-1}\hat{u} = D^k u_0$ with $u_0 \in S'_0$.[4]

PROOF. The statement of the theorem together with the S' representation Theorem 1.28 allows one to write, for some k, $\hat{u} = (i\omega)^k \hat{g}$, where $\hat{g} = \mathscr{F}[g \in \mathscr{D}'_+] \in \mathscr{D}'_{L_2}$. Thus, by Theorem 1.36, and noting that $\mathscr{F}H(t) = \pi\delta + \text{pv}(1/i\omega)$, we obtain $\hat{g} = \mathscr{F}[gH(t)] = (1/2\pi)\hat{g} * [\pi\delta + \text{pv}(1/i\omega)]$, which is defined because $\text{pv}(1/\omega) \in \mathscr{D}'_{L_2}$. Solving for $\hat{g}$ yields $\hat{g} = (1/\pi i)\hat{g} * \text{pv}(1/\omega)$. Therefore

$$D^k \hat{u} = D^k[(i\omega)^k \hat{g}] = D^k \left\{ \frac{(i\omega)^k}{\pi i} \left[\hat{g} * \text{pv} \frac{1}{\omega} \right] \right\}$$

which may be written, in view of Lemma 3.1, as

$$D^k \hat{u} = \frac{1}{\pi i} [(i\omega)^k \hat{g}] * D^k \text{pv} \frac{1}{\omega} = \frac{1}{\pi i} \hat{u} * D^k \text{pv} \frac{1}{\omega}.$$

The alternate condition may be established by terminating the discussion at the point $\hat{g} = (1/\pi i)\hat{g} * \mathrm{pv}(1/\omega)$ since $(i\omega)^k\hat{g} = \hat{u}$. Conversely, given (3.15), Lemma 3.1 implies that

$$D^k\hat{u} = D^k[(i\omega)^k\hat{g}] = D^k\left\{\frac{(i\omega)^k}{\pi i}\left[\hat{g} * \mathrm{pv}\frac{1}{\omega}\right]\right\}$$

or $D^k\{(i\omega)^k[\hat{g} - (1/\pi i)\hat{g} * \mathrm{pv}(1/\omega)]\} = 0$. The expression within the braces is in S' so we may apply Fourier transforms to obtain

$$t^k\{\mathfrak{F}^{-1}(i\omega)^k[\quad]\} = t^k\{D^k\mathfrak{F}^{-1}[\quad]\} = 2t^k\{D^k[g - gH(t)]\} = 0,$$

from which we conclude that

$$D^k g = \mathfrak{F}^{-1}\hat{u}_\omega = D^k[gH(t)] + \sum_{j \leq k-1} a_j D^j \delta \in \mathscr{D}'_+.$$

Similarly, starting with (3.16), we may follow essentially the same argument.

The conditions of the theorem, either (3.15) or (3.16), may be considered as generalized Hilbert pairs since by considering the real and imaginary components of $\hat{u}_\omega$ we may separate (3.15) into

$$D^k \operatorname{Re} \hat{u}_\omega = \frac{1}{\pi}(\operatorname{Im}\hat{u}) * D^k \mathrm{pv}\frac{1}{\omega} \tag{3.17}$$

and

$$D^k \operatorname{Im} \hat{u}_\omega = -\frac{1}{\pi}(\operatorname{Re}\hat{u}) * D^k \mathrm{pv}\frac{1}{\omega}. \tag{3.18}$$

Of course, the result takes on a more recognizable form where $k = 0$, i.e., $\operatorname{Re}\hat{u} = (1/\pi)\operatorname{Im}\hat{u} * \mathrm{pv}(1/\omega)$, and $\operatorname{Im}\hat{u} = -(1/\pi)\operatorname{Re}\hat{u} * \mathrm{pv}(1/\omega)$. However, $k = 0$ corresponds to $\hat{u}_\omega \in \mathscr{D}'_{L_2}$ and we realize again the natural extension of the L_2 classic results to the $\mathscr{D}'_{L_2}$ distributional statements.

The previous theorem is central in the study of distributional boundary values because it reflects the holomorphic character of the half-plane function $\hat{u}(p)$ into a precise distributional statement involving only its distributional boundary value $\hat{u}_\omega$. The result is a concise statement which we will have occasion to use in future sections. However, at this point we would like to remark that the intimate bond, displayed by (3.8), between the real and imaginary components of the boundary values has wide physical interest in the study of quantum theory in which context it is referred to as a dispersion relation. Indeed, since the quantum system under study may be represented by linear operators, the dispersion relations, as we realize on the basis of Section 3.1, reflect the causal property of such operators.[5] One might also note that the generalized Hilbert pairs (3.15) and (3.16) are in fact the distributional limits of the two alternate Cauchy integrals (3.8) and (3.9) which were obtained in Theorem 3.4. This complete symmetry is of course not accidental and in fact was sought after on the basis of the form of the classic L_2 results. Thus, for example, if $k = 0$ and if $f(p) = 1/2\pi \langle f_\zeta, 1/(p - i\zeta) \rangle$ then $f(p) \xrightarrow{S} f_\omega$ and $1/2\pi \langle f_\zeta, 1/(p - i\zeta) \rangle \to (1/2\pi) f_\zeta * [\pi\delta + \text{pv}(1/i\omega)]$ from which we obtain $f_\omega = (f_\omega/2) + (1/2\pi i) f_\omega * \text{pv} \, 1/\omega$ and so $f_\omega = (1/\pi i) f_\omega * \text{pv} \, 1/\omega$. (Note that the Cauchy integral converges to formula (3.12) from the right (with $k = 0$), and this formula is equivalent to the Hilbert transform only when $\mathfrak{F}^{-1} f_\omega \in \mathcal{D}'_+$.)

It may be of interest to note that the direct extension of the classic theory of Hilbert transforms may be developed quite independently from the previous study. Thus if we define the Hilbert transform $\overset{v}{g}$ of g as

$$\overset{v}{g} = \frac{1}{\pi} g * \text{pv} \frac{1}{\omega},$$

then the inverse transform, or Hilbert pair, is obtained as

$$g = -\frac{1}{\pi} \overset{v}{g} * \text{pv} \frac{1}{\omega}.$$

The formal theory may be summarized in

Theorem 3.11. Let $g \in \mathcal{D}'_{L_2}$. Then $\overset{v}{g}$ exists and is contained $\mathcal{D}'_{L_2}$. Moreover,

$$g = -\frac{1}{\pi} \overset{v}{g} * \text{pv} \frac{1}{\omega}.$$

It should be noted that although Hilbert transform pairs exist for any $g \in \mathscr{D}'_{L_2}$ it does not follow that $-\operatorname{Re} \overset{v}{g} = \operatorname{Im} g$ or $\operatorname{Im} \overset{v}{g} = \operatorname{Re} g$. In fact, such reciprocal relations for the real and imaginary pairs will hold when and only when the conditions of Theorem 3.10 are satisfied.

PROOF. $g \in \mathscr{D}'_{L_2}$ may be represented as $g = \Sigma_j D^j h_j$ with $h_j \in L_2$. Thus

$$\overset{v}{g} = \frac{1}{\pi} \left(\sum_j D^j h_j \right) * \operatorname{pv} \frac{1}{\omega} = \frac{1}{\pi} \sum_j D^j \left[h_j * \operatorname{pv} \frac{1}{\omega} \right] \in \mathscr{D}'_{L_2}$$

because $h_j * \operatorname{pv}(1/\omega) \in L_2$ for $h_j \in L_2$ [T1, Chapter 5]. The reciprocal relationship is established by noting that $-(1/\pi)\overset{v}{g} * \operatorname{pv}(1/\omega) = -(1/\pi)\{[(1/\pi)g * \operatorname{pv}(1/\omega)] * \operatorname{pv}(1/\omega)\}$ and $\operatorname{pv}(1/\omega) * \operatorname{pv}(1/\omega) = -\pi\delta$ as we may demonstrate by applying Fourier transforms with

$$\mathfrak{F}^{-1}[\operatorname{pv}(1/\omega)] = (i/2) \operatorname{sgn} t.$$

This theorem in its classic setting was restricted to $g \in L_2$. In the distributional form, however, it may be established for $g \in \mathscr{D}'_{L_p}$ with $\overset{v}{g} \in \mathscr{D}'_{L_p}$ for $p > 1$ and for $g \in \mathscr{D}'_{L_1}$ in which case $\overset{v}{g} \in \mathscr{D}'_{L_r}$ $r > 1$. In either case g may be obtained from $\overset{v}{g}$ by means of the inverse transform.

At this point we can summarize many of the results of this and the preceding section in a way that illustrates their connection with the classic work of Paley and Wiener [P1], Titchmarsh [T1], and Hille and Tamarkin [H1]. In particular, the following theorem is appropriate

Theorem 3.12. The necessary and sufficient conditions that $\hat{u}(\omega) \in L_2$ be the boundary value, in the L_2 norm, of a function $u(p)$ holomorphic in $\operatorname{Re} p > 0$ satisfying $\sup_{\sigma > 0} \int_{-\infty}^{\infty} |u(\sigma + i\omega)|^2 \, d\omega < \infty$, i.e., $u(p) \in H^2$, is either

(1) $u = \mathfrak{F}^{-1}\hat{u} \in L_2(0, \infty)$,
(2) $C(\hat{u}, p) = \mathscr{L}u, \operatorname{Re} p > 0$,
(3) $C(\hat{u}, p) = 0 \operatorname{Re} p < 0$, i.e., $P(\hat{u}, p) = C(\hat{u}, p)$ in $\operatorname{Re} p > 0$, or
(4) $\hat{u} = (1/\pi i)\hat{u} * \operatorname{pv}(1/\omega)$.

The distributional version of the theorem has been established in the foregoing discussion and can be stated in the same manner as Theorem 3.12 if we replace $\hat{u}(\omega) \in L_2$ by $\hat{u}_\omega \in \mathscr{D}'_{L_2}$, and the L_2 norm by S'

convergence. In view of this comparison one would conjecture that $\mathscr{D}'_{L_2}$ is the natural generalization, at least in the area of boundary value studies, of the L_2 class. It is noteworthy that although the classic theorem stated above is valid for $u \in L_p, 1 < p \leq 2$, the distributional version holds for $u \in \mathscr{D}'_{L_p}, 1 \leq p \leq 2$ so that, in particular, if $u \in L_1 \subset \mathscr{D}'_{L_1}$ the theorem is applicable but the statements must be interpreted, in certain cases, in a distributional sense. The most general statement of the preceding theorem is of course obtained when $\hat{u}(\omega) \in S'$. We summarize many of the preceding results in the following theorem.

Theorem 3.13. The necessary and sufficient conditions that $\hat{u}_\omega \in S'$ be the boundary value, in the S' topology, of a function $\hat{u}(p) \in H^+$ is either

(1) $u = \mathfrak{F}^{-1}\hat{u}_\omega \in S' \cap D'_+$,
(2) $p^k C[(\hat{u}_\omega/(i\omega)^k), p] = \hat{u}(p) = \mathscr{L}u, \operatorname{Re} p > 0.$
(3) $C[(\hat{u}_\omega/(i\omega)^k), p = 0, \operatorname{Re} p < 0$, i.e., $p^k P[(\hat{u}_\omega/(i\omega)^k), p]$
  $= p^k C[(\hat{u}_\omega/(i\omega)^k), p]$ in $\operatorname{Re} p > 0$, or
(4) $D^k \hat{u} = (1/\pi i)\hat{u} * D^k \operatorname{pv}(1/\omega).$

Before closing this section we pause again in order to briefly summarize the reasons for studying the boundary values of holomorphic functions in a distributional setting, and, more directly, what kind of results one can expect from such a study. In the first place, if $\hat{f}(p) \in H^+$, then $\lim_{\sigma \to 0} \hat{f}(\sigma + i\omega)$ can occur in a variety of ways, if it exists at all. If the L_q norm limit does not exist, it is relevant to ask how intimate a relationship or bond obtains between the holomorphic function and its boundary value. One answer, of course, lies in the results of the distributional version of Theorem 3.12 and, more generally, in Theorem 3.13. We should emphasize in this regard that the more we relax the allowable mode of convergence to the boundary (e.g., norm convergence implies distributional or weak convergence but not conversely), the larger an admissible class of boundary values can be allowed and, correspondingly, we can consider a larger class of holomorphic functions. Thus if we insist on applying Theorem 3.12 in its classic setting, then even the simple example $1/p$ will convince one that no classic connection exists between the example and its point-wise a.e. boundary value. However, the S' boundary value can reproduce uniquely the half-plane function by means, for example, of the generalized Cauchy integral. In addition, as we will demonstrate later in this chapter, certain situations exist in which one can make more precise statements on the boundary, or one can display most vividly the inherent erratic behavior of the holomorphic function, by working directly on the boundary.

3.4 ANALYTIC CONTINUATION AND UNIQUENESS

One of the central questions in the theory of analytic functions concerns the extension of holomorphic functions to some larger regions than their original domain of definition. Basically, these problems require some means of definition in the enlarged domain and some assurance that the continuation is unique. The classic result of Painleve [Pa1] concerning this question may be phrased in the following manner: given two functions such that the boundaries of their domain of holomorphy share a common arc, and assume that the two holomorphic functions take on the same boundary values on this arc in a continuous manner from within each domain respectively; then the two functions are analytic continuations of each other in the sense that there exists a single function equal to the two functions in the separate domains, and this single function is holomorphic in the enlarged domain composed of the original two and the common arc of the boundaries. We will describe an extension of this result in which the continuity requirement is relaxed to a weak distributional convergence and the equality of the two separate boundary values is to be regarded in a distributional sense. The remarkable statement that the two functions are analytic continuations of each other and yield a holomorphic function on the arc is indicative of the generality of the distributional approach. We will content ourselves with a proof of a restricted version of the more general theorem. Specifically, we will consider the domains to be half planes.

Theorem 3.14. Let $u^+(p)$ and $u^-(p)$ be H^+ functions in the respective planes $\operatorname{Re} p > 0$ and $\operatorname{Re} p < 0$, and assume that they separately converge in the S' topology to the distributional boundary values u_ω^+ and $u_\omega^- \in S'$ such that $\langle (u^+ - u^-), \varphi \rangle = 0$ for all $\varphi \in C_0^\infty$ whose support is contained in some open set U of the boundary. Then there exists a single function $u(p)$ that is equal to $u^+(p)$ in $\operatorname{Re} p > 0$, and $u^-(p)$ in $\operatorname{Re} p < 0$ and, in addition, is holomorphic in the entire plane with the exception of that portion of the line $p = i\omega$ which corresponds to the complement of U.[6]

PROOF. Since open sets are unions of open intervals, we will obtain the stated result when U is an open interval; the extension to open sets follows immediately. Thus, employing the representation of Theorem 3.4 we obtain for some n (although in a direct application of the theorem, two unequal n's may be used, they can always be made equal to the larger of the two since the S' representation theorem, which is the basis for the theorem, will maintain its form for all n larger than some minimum integer) $u^+(p) = p^n C(u_\omega^+/(i\omega)^n, p)$, $\operatorname{Re} p > 0$, and $u^-(p) = -p^n C(u_\omega^-/(i\omega)^n, p)$, $\operatorname{Re} p < 0$. Now, by Corollary 3.1, $C(u^+/(i\omega)^n, p) = 0$, $\operatorname{Re} p < 0$, and $C(u^-/(i\omega)^n, p) = 0$, $\operatorname{Re} p > 0$, so if we define $u(p) = p^n[C(u^+/(i\omega)^n, p) - C(u^-/(i\omega)^n, p)]$ we have $u(p) = u^+(p)$ for $\operatorname{Re} p > 0$ and $u(p) = u^-(p)$ for $\operatorname{Re} p < 0$. In order to conclude the proof, we must establish that $u(p)$ is holomorphic on the open interval U since $u(p)$ by its definition is holomorphic in the separate half planes. To this end define $\lambda(\zeta) \in C^\infty$ equal to 0 on some arbitrary closed interval $U' \subset U$ and one on the complement of U (see Lemma 1.3). Note that since U' is arbitrary, the distance between it and U can be made arbitrarily small. Then since $\langle (u^+ - u^-), \varphi \rangle = 0$ for all $\varphi \in C_0^\infty$ whose support is contained in the open set U, it will remain zero when the φ have their support in U'. Thus $u(p) = p^n \langle (u^+/(i\omega)^n - u^-/(i\omega)^n),$ $[\lambda(\zeta)/(p - i\zeta)] \rangle$. But $\{[\lambda(\zeta)/(p' - i\zeta)] - [\lambda(\zeta)/(p - i\zeta)]\}1/(p' - p)$ tends to $-[\lambda(\zeta)/(p - i\zeta)^2]$ in the $\mathscr{D}_{L_2}$ topology as $p' \to p$ for all $p \in U'$. Thus employing the same reasoning as used in the proof of Theorem 3.2, leads to the conclusion that $du(p)/dp$ exists for $p \in U'$ which may be made arbitrarily close to U.

A rather immediate consequence of this theorem is that if U is the entire axis then the analytic continuation is an entire function, viz. a polynomial.

As we noted previously, this theorem is but a special case of a more general theorem in which the two domains need not be half planes and in which the topology need only be that of $\mathcal{D}'$. However, the proof of this result requires concepts which we have not developed and so we will accept Theorem 3.14, and its proof, as indicative of the type of theorem possible where the distributional feaures of the problem are considered (cf. Appendix II).

Another aspect of the uniqueness of holomorphic functions is involved when one considers the uniqueness of their distributional boundary values. Let us assume that two H^+ functions converge in the S' topology to distributional boundary values that are identical on some open set, the immediate question is whether or not the two functions are identical. Assuming that they were not equal would lead to the conclusion (by defining a new function to be the difference of the two) that a nonzero H^+ function that converges in the S' topology could have a distributional boundary value equivalent to zero on some open set. Since the following theorem proves that this is impossible, we conclude that the distributional boundary value on any open set uniquely determines not only the originating holomorphic function but the remainder of the distributional boundary value on the complement of the set. Thus we wish to establish the following corollary to Theorem 3.14.

Corollary 3.4. If $u_\omega \in S'$ is the S' boundary value of some H^+ function then u_ω cannot vanish on any open set, unless u_ω vanishes identically.

PROOF. As with the proof of Theorem 3.14 we need only consider open intervals. Thus we assume that $u(p) \xrightarrow{S'} u_\omega$ and u_ω vanishes in some open interval. But the function 0 is holomorphic in Re $p < 0$ and converges in the S' topology to zero in Re $p < 0$, i.e., its boundary value is zero in the open interval. The analytic continuation principle of Theorem 3.13 implies that $u(p)$ is the unique continuation of the function 0 which, however, is trivially known to be the function 0 in the entire plane, i.e., $u(p)$ must be zero in $\operatorname{Re} p > 0$ and so its S' boundary value must also be zero.

The corollary and the preceding discussion of the uniqueness of distributional boundary values raises the following question. Under what conditions will a distribution that is known only in some open set of the $p = i\omega$ axis, uniquely determine an H^+ function that has S' boundary values and, in particular, the same boundary value in the open set. We will obtain an algorithm that determines when this "continuation" away from the boundary is possible. Specifically, we consider distributional boundary values $u_\omega \in \mathscr{D}'_{L_2}$, which are known on some open interval U. Note that since $u_\omega \in \mathscr{D}'_{L_2}$, $u_\omega = \Sigma_j D^j u_j$ where the $u_j \in L_2$. If we define $\alpha(\zeta)$ to be one on any arbitrary closed interval $K' \supset U$ and zero on the complement of U, then $\langle \Sigma_j (D^j[u_j\alpha] - D^j u_j), \varphi \rangle = 0$ for all φ whose support is contained in K'. By this artifice, we have decomposed the boundary distribution u_ω (if it exists) into two well defined distributions $\Sigma_j D^j(u_j\alpha)$, and $u_\omega - \Sigma_j D^j(u_j\alpha)$ with the property that $u_\omega - \Sigma_j D^j(u_j\alpha)$ must vanish, distributionally, on the set K'. Again, if we assume that there exists a u_ω which satisfies the demands of this problem $\mathfrak{F}^{-1}u_\omega = \mathfrak{F}^{-1}[\Sigma_j D^j(u_j\alpha)] + f \in \mathscr{D}'_+$ where f is the inverse Fourier transform of $u_\omega - \Sigma_j D^j(u_j\alpha)$. But since it was assumed that $u_\omega \in \mathscr{D}'_{L_2}$, both f and $\mathfrak{F}^{-1}[\Sigma_j D^j(u_j\alpha)]$ are tempered locally integrable functions. Thus if $\mathfrak{F}^{-1}u_\omega \in \mathscr{D}'_+$ it is necessary that

$$\mathfrak{F}^{-1}[\Sigma_j D^j(u_j\alpha)] + f = 0$$

pointwise, a.e. for $t < 0$, or equivalently $H(-t)\mathfrak{F}^{-1}[\Sigma_j D^j(u_j\alpha)] = -H(-t)f$. But $\mathfrak{F}f = \mathfrak{F}[fH(t)] + \mathfrak{F}[fH(-t)]$ and since $\mathfrak{F}f$ must vanish on K' we obtain $\langle (\mathfrak{F}[fH] + \mathfrak{F}[fH(-t)]), \varphi \rangle = \langle (\mathfrak{F}[fH] - \mathfrak{F}\{H(-t)\mathfrak{F}^{-1}[\Sigma_j D^j(u_j\alpha)]\}), \varphi \rangle = 0$ for all $\varphi \in \mathscr{D}$ whose support is contained in K'. Appealing to the analytic continuation principle of Theorem 3.13, we conclude that $\mathscr{L}[fH(t)]$ must be the continuation of the function $\mathscr{L}\{H(-t)\mathfrak{F}^{-1}[\Sigma_j D^j(u_j\alpha)]\}$ that is contained in H^+ for $\operatorname{Re} p < 0$. If the continuation is possible, we will have determined $\mathscr{L}[fH(t)]$ of $fH(t)$ itself so that f would be completely determined since

$\mathfrak{F}^{-1}[\Sigma_j D^j(u_j\alpha)] = -f$ for $t < 0$. The entire distribution u_ω would then be obtained directly by computing $u_\omega = \mathfrak{F}f + \Sigma_j D^j(u_j\alpha)$. The algorithm then involves determining whether the H^+ function, $\mathcal{L}\{\mathfrak{F}^{-1}[\Sigma_j D^j(u_j\alpha)]H(-t)\}$, which is obtained directly from the known distribution, can be analytically continued across the open interval U into the entire half plane $\operatorname{Re} p > 0$ such that the continued function has a $\mathcal{D}'_{L_2}$ boundary value in $\operatorname{Re} p > 0$. Note that the continuation process may proceed in principle by the classic chain-of-circles technique employing power series expansions; we imply that if the continuation can be obtained, it must be possible in a classic manner since the continued function must be holomorphic on the open interval U. Although the above algorithm is obviously not a complete solution to the stated problem, it yields at least one avenue of approach.

3.5   PASSIVE OPERATORS[7]

In such diverse areas as the study of nuclear particle scattering, the theory of electromagnetic wave scattering, and the study of electrical networks, the concept of a passive or dissipative operator has proven to be of great utility. The aim of this section is to apply some of the results of the previous section to a study of such operators in order to obtain concise representations of them. Of the many alternative definitions, we will consider first the following definition.

DEFINITION 3.3.   A linear, continuous, translation invariant, causal operator mapping real $\mathbf{u} \in \mathscr{E}'$ into real $\mathbf{v} \in \mathscr{D}'$ is said to be passive if $\int_{-\infty}^{t} \mathbf{v}^T\mathbf{u}\, dx \geq 0$ for all real $\mathbf{u} \in C_0^\infty$ and all t. It follows easily, from the linearity of the operator that this definition is equivalent to requiring that

$$\int_{-\infty}^{t} (\bar{\mathbf{v}}^T\mathbf{u} + \bar{\mathbf{u}}^T\mathbf{v})\, dx \geq 0 \qquad (3.19)$$

for all $\mathbf{u} \in C_0^\infty$. The notation $\mathbf{f}$ indicates an $n \times 1$ matrix of elements, and T denotes transpose. For convenience, the operator will be referred to as a passive immittance operator.

It may be noted that since the operator maps real distributions into real distributions, the definition is equivalent to

$$2 \operatorname{Re} \int_{-\infty}^{t} \bar{\mathbf{v}}^{T} \mathbf{u} \, dx \geq 0 \qquad (3.20)$$

when $\mathbf{u}$ is not real. The latter expression has the physical significance in an electric network, where $\mathbf{u}$ may denote the voltages applied to the system and $\mathbf{v}$ the resulting currents, of being the net energy delivered to the network. The implication of the definition is that the network is capable of absorbing (or storing) electrical energy but not generating it.[8] We will obtain a characterization of such passive operators following the proof developed by Zemanian [Z2]. Initially, we consider the following two lemmas.

Lemma 3.2. The generalized Green's function $w(t)$ (an $n \times n$ matrix of real distributions) of a passive immittance operator is contained in $\mathscr{D}'_{L_{\infty}}$, i.e., each of the elements of $w(t)$ is contained in $\mathscr{D}'_{L_{\infty}}$.

PROOF. Let $\mathbf{u} = \mathbf{y}\varphi$ where $\mathbf{y}$ is a constant $n \times 1$ vector and $\varphi \in C_0^{\infty}$. The convolution representation of the operator (being linear, continuous, and translation invariant) implies that $\mathbf{v} = w(t) * \varphi \mathbf{y}$ where the $*$ indicates a scalar convolution operation (Theorem 1.18 may be easily extended to vector-valued distributions and, in this case, $T\delta$ is a matrix-valued distribution. The notation $A * u$, when A is a matrix-valued distribution and u is a scalar, is to be interpreted in a term by term manner.) The passivity of the operator may be phrased for $t \to \infty$ as $\int_{-\infty}^{\infty} [\bar{\mathbf{y}}^{T} w^{T} * \bar{\varphi}][\mathbf{y}\varphi] + [\bar{\mathbf{y}}^{T}\bar{\varphi}][w * \varphi \mathbf{y}] \, dx = \int_{-\infty}^{\infty} [(\bar{\mathbf{y}}^{T} w^{T} \mathbf{y}) * \bar{\varphi}]\varphi + [(\bar{\mathbf{y}}^{T} w \mathbf{y}) * \varphi]\bar{\varphi} \, dx \geq 0$. The following identities are valid when either u or $v \in C_0^{\infty}$: $\int_{-\infty}^{\infty} uv \, dx = \langle \delta, u * \tilde{v} \rangle = \langle \delta, \tilde{u} * v \rangle$, where $\tilde{k}(t) = k(-t)$ and, distributionally, $\langle \tilde{u}, \varphi \rangle = \langle u, \tilde{\varphi} \rangle$. Thus we may express the passivity condition at infinity as

$$\int_{-\infty}^{\infty} [(\bar{\mathbf{y}}^{T} w^{T} \mathbf{y}) * \bar{\varphi}]\varphi + [(\bar{\mathbf{y}}^{T} w \mathbf{y}) * \varphi]\bar{\varphi} \, dx =$$

$$\langle \delta, [(\bar{\mathbf{y}}^{T} w^{T} \mathbf{y}) * \bar{\varphi} * \tilde{\varphi}] \rangle + \langle \delta, \widetilde{[(\bar{\mathbf{y}}^{T} w \mathbf{y}) * \varphi]} * \bar{\varphi} \rangle \geq 0.$$

However,

$$[(\bar{\mathbf{y}}^T w^T \mathbf{y}) * \varphi] = (\bar{\mathbf{y}}^T \tilde{w}^T \mathbf{y}) * \tilde{\varphi},$$

and we finally obtain

$$\langle \delta, [\bar{\mathbf{y}}^T (w + \tilde{w}^T) \mathbf{y}] * (\bar{\varphi} * \tilde{\varphi}) \rangle = \langle [\bar{\mathbf{y}}^T (w + \tilde{w}^T) \mathbf{y}], \varphi * \bar{\tilde{\varphi}} \rangle \geq 0$$

for all $\varphi \in C_0^\infty$. If $\langle \chi, \varphi * \bar{\tilde{\varphi}} \rangle \geq 0$ for all $\varphi \in C_0^\infty$ then χ is said to be a nonnegative definite distribution [Sc1]. Now for any nonnegative definite distribution χ note that

$$\langle (\chi * \varphi * \bar{\tilde{\varphi}}), \bar{\varphi} * \tilde{\varphi} \rangle = \langle \chi_y, \langle \varphi * \bar{\tilde{\varphi}}, (\bar{\varphi} * \tilde{\varphi})_{x+y} \rangle \rangle =$$

$$\langle \chi_y, [(\varphi * \bar{\tilde{\varphi}}) * (\bar{\tilde{\varphi}} * \varphi)] \rangle = \langle \chi_y, (\tilde{\varphi} * \bar{\varphi}) * (\bar{\varphi} * \tilde{\varphi}) \rangle \geq 0$$

because with $\varphi \in C_0^\infty$, $\bar{\varphi} * \tilde{\varphi} \in C_0^\infty$. However, since $\chi * \varphi * \bar{\tilde{\varphi}}$ is a continuous function, and we have just established that it is nonnegative definite, we may conclude, as is well known classically, that it must be bounded. Finally, by means of the identity $4(\alpha * \beta) = (\alpha + \bar{\beta}) * (\bar{\alpha} + \beta) - (\alpha - \bar{\beta}) * (\bar{\alpha} - \beta) + i(\alpha + i\bar{\beta}) * (\bar{\alpha} - i\beta) - i(\alpha - i\bar{\beta}) * (\bar{\alpha} + i\beta)$ we conclude that $\chi * \alpha * \beta$ must be bounded when χ is nonnegative definite or by repeated application of Theorem 1.32 $\chi \in \mathscr{D}'_{L_\infty}$. Thus we have established that $\bar{\mathbf{y}}^T (w + \tilde{w}^T) \mathbf{y} \in \mathscr{D}'_{L_\infty}$ for all constant vectors $\mathbf{y}$. If we select a $\mathbf{y}$ such that all its elements are zero except for y_r and y_k we obtain

$$\bar{\mathbf{y}}^T (w + \tilde{w}^T) \mathbf{y} = y_r \bar{y}_r [w_{rr} + \tilde{w}_{rr}] + \bar{y}_r y_k [w_{rk} + \tilde{w}_{kr}] + \bar{y}_k y_r [w_{kr} + \tilde{w}_{rk}]$$

$$+ \bar{y}_k y_k [w_{kk} + \tilde{w}_{kk}] \in \mathscr{D}'_{L_\infty}.$$

Setting $y_k = 0$ or $y_r = 0$ yields $w_{rr} + \bar{w}_{rr} \in \mathscr{D}'_{L_\infty}$ and $w_{kk} + \tilde{w}_{kk} \in \mathscr{D}'_{L_\infty}$. Therefore $\bar{y}_r y_k [w_{rk} + \tilde{w}_{kr}] + \bar{y}_k y_r [w_{kr} + \tilde{w}_{rk}] \in \mathscr{D}'_{L_\infty}$ and if $y_r = y_k = 1, w_{rk} + \tilde{w}_{rk} + w_{kr} + \tilde{w}_{kr} \in \mathscr{D}'_{L_\infty}$, whereas if $y_r = 1$, $y_k = i$, $w_{rk} - \tilde{w}_{rk} - w_{kr} + \tilde{w}_{kr} \in \mathscr{D}'_{L_\infty}$. Adding the latter two statements yields $w_{rk} + \tilde{w}_{kr} \in \mathscr{D}'_{L_\infty}$. However, since each element of the matrix $\in \mathscr{D}'_+$ (the operator is causal), the sum, for example $w_{rk} + \tilde{w}_{kr}$, reflect the entire behavior of the distribution except possibly for a distribution of point support at the origin which may appear in both terms and cancel. But since distributions of point support belong to $\mathscr{D}'_{L_\infty}$, we conclude finally that the four elements $w_{rr}, w_{kr}, w_{kk}, w_{rk} \in \mathscr{D}'_{L_\infty}$. By this means we may consider each element of the w matrix and arrive at the same conclusion.

Lemma 3.3. A passive immittance operator yields

$$\int_{-\infty}^{t} [\bar{\mathbf{v}}^T \mathbf{u} + \bar{\mathbf{u}}^T \mathbf{v}] \, dx \geq 0$$

for all $\mathbf{u} \in S$ and all t.

PROOF. Lemma 3.2 established the fact that $w(t) \in \mathscr{D}'_{L_\infty}$ so that $\mathbf{v} = w * \mathbf{u}$ is defined for all $\mathbf{u} \in S$. However, C_0^∞ is dense in S so that a sequence $\mathbf{u}_n \in C_0^\infty \to \mathbf{u} \in S$ yields, due to the continuity of the convolution representation of the operator, $\mathbf{v}_n \to \mathbf{v}$. Moreover, for each n, the passivity integral is nonnegative so that if the sequence of integrals converge, the result must be nonnegative. However, $\mathbf{v}_n$ are all uniformly bounded functions since $w \in \mathscr{D}'_{L_\infty}$. The bounded convergence theorem and simple estimates then establishes that the integrals converge.

We are now in a position to establish a characterization of passive immittance operators.

Theorem 3.15. Let $w(t)$ be the generalized Green's function of a passive immittance operator (Definition 3.3), then its Laplace transform $W(p)$ is such that for Re $p > 0$

(1) $W(p)$ is holomorphic,

(2) $W(\bar{p}) = \overline{W(p)}$,

(3) $\overline{W(p)}^T + W(p)$ is a nonnegative definite matrix. Such a matrix $W(p)$ is said to be positive-real.

PROOF. Let $\mathbf{u} = \mathbf{y} \, e^{pt} \lambda(t)$ with Re $p > 0$, $\mathbf{y}$ a constant $n \times 1$ vector, and $\lambda(t) \in C^\infty$ is one for $t \leq t_0$ and zero for $t \geq t_0 + \varepsilon$. Then $\mathbf{u} \in S$ and in particular $\mathbf{u} = \mathbf{y} \, e^{pt}$ for $t \leq t_0$. One may now compute the $\mathbf{v}$ corresponding to $\mathbf{u} = \mathbf{y} \, e^{pt}$ for Re $p > 0$ as $\mathbf{v} = w * \mathbf{y} \, e^{pt} = e^{pt} W(p)\mathbf{y}$, where $W(p)$ is the Laplace transform of $w \in \mathscr{D}'_+$ i.e., for all $\varphi \in C_0^\infty$, $\langle w * \mathbf{y} \, e^{pt}, \varphi \rangle = \langle w\mathbf{y}, \langle e^{px}, \varphi(t+x) \rangle \rangle = \langle w\mathbf{y}, \bar{e}^{pt} \rangle \langle e^{px}, \varphi(x) \rangle = \langle W(p)\mathbf{y} \, e^{px}, \varphi(x) \rangle$. $W(p)$ will be holomorphic in Re $p > 0$ because

$w(t) \in \mathscr{D}'_{L_\infty} \cap \mathscr{D}'_+$ (the operator is causal; Lemma 3.2). Again, since $w(t) \in \mathscr{D}'_+$, $\mathbf{v} = w * (\mathbf{y} \, e^{pt} \lambda(t)) = w * (\mathbf{y} \, e^{pt}) = e^{pt} W(p)\mathbf{y}, t \leq t_0$, and therefore for $\mathrm{Re}\, p > 0$,

$$\int_{-\infty}^{t_0} (\bar{\mathbf{v}}^T \mathbf{u} + \bar{\mathbf{u}}^T \mathbf{v}) \, dx = \bar{\mathbf{y}}^T [\overline{W(p)}^T + W(p)]\mathbf{y} \int_{-\infty}^{t_0} e^{2\mathrm{Re}\, pt} \, dt \geq 0,$$

which implies that $\bar{\mathbf{y}}^T [\overline{W(p)}^T + W]\mathbf{y} \geq 0$ or $\overline{W(p)}^T + W(p)$ must be nonnegative definite in $\mathrm{Re}\, p > 0$. Finally, since each element of the matrix $w(t)$ is real, $W(\bar{p}) = \overline{W(p)}$.

The converse of the theorem, that every positive-real matrix represents a passive immittance operator, has been established by Zemanian [Z2] but the details of this proof would take us too far afield. In the context of the previous investigations of the distributional boundary behavior of holomorphic functions, we focus attention instead on the boundary values of positive-real functions. This will lead us to a very concise characterization of these functions and therefore to a characterization of passive immittance operators. The discussion will employ the following classic representation theorem due to Youla [Yo2], a proof of which is given in Appendix I.

Theorem 3.16. The necessary and sufficient condition that an $n \times n$ matrix $W(p)$ be positive-real is that for $\mathrm{Re}\, p > 0$,

$$W(p) = Q + Ap + \int_{-\infty}^{\infty} \frac{1 - pi\tau}{p - i\tau} \, dM(\tau), \qquad (3.21)$$

where Q is a real, constant, skew-symmetric $n \times n$ matrix $(Q^T = -Q)$; A is a real, constant, symmetric, nonnegative definite matrix; and M is an $n \times n$ matrix each of whose elements are of bounded variation on the entire line $-\infty < \tau < \infty$, satisfying $\overline{M}^T = M$, $-\overline{M(-\tau)} = M(\tau)$ and with $\bar{\mathbf{a}}^T M(\tau)\mathbf{a}$ being a real, bounded nondecreasing function of τ for any complex constant $n \times 1$ vector $\mathbf{a}$.

By employing Youla's theorem, we will establish the following theorem.

Theorem 3.17. The necessary and sufficient condition that an $n \times n$ matrix of distributions W_ω be the boundary behavior in the S' topology of a positive real matrix is that

$$W_\omega = Q + i\omega\left[A + \int_{-\infty}^{\infty} dM(\tau)\right] + \pi(1 + \omega^2)\,DM(\omega)$$

$$+ (1 + \omega^2)\left[DM * \mathrm{pv}\,\frac{1}{i\omega}\right], \tag{3.22}$$

where Q, A, and M are defined in Theorem 3.15. In particular,

$$\overline{W}_\omega^T + W_\omega = 2\pi(1 + \omega^2)\,DM(\omega). \tag{3.23}$$

PROOF. The necessity will be established if we demonstrate that condition (3.22) of the theorem is the S' boundary value of the representation for $W(p)$ given in Theorem 3.16. The first two terms in $W(p)$ converge in their present form with $p \to i\omega$. The third term may be rewritten, by noting that $[(1 - pi\tau)/(p - i\tau)] = p + [(1 - p^2)/(p - i\tau)]$, in the form $p\int_{-\infty}^{\infty} dM + (1 - p^2)\int_{-\infty}^{\infty}[dM/(p - i\tau)]$, where we have observed that the bounded variation of $M(\tau)$ implies that $\int_{-\infty}^{\infty} dM$ exists. The third term of (3.22) is then obtained directly. Finally, consider, for all $\varphi \in S$,

$$\left\langle (1 - p^2)\int_{-\infty}^{\infty} \frac{dM}{p - i\tau}, \varphi(\omega)\right\rangle = \int_{-\infty}^{\infty} dM(\tau)\int_{-\infty}^{\infty} \frac{(1 - p^2)\varphi(\omega)}{p - i\tau}\,d\omega,$$

where the classic Fubini theorem justifies the interchange. However, Parseval's relation implies that

$$\int_{-\infty}^{\infty} \frac{(1 - p^2)\varphi(\omega)}{p - i\tau}\,d\omega = 2\pi\int_{-\infty}^{\infty} \hat{\varphi}_\sigma(t)\,e^{i\tau t}\,e^{-\sigma t}H(t)\,dt,$$

where $\hat{\varphi}_\sigma(-t) = \mathfrak{F}^{-1}[(1 - p^2)\varphi(\omega)] \in S$ for every $\sigma \geq 0$, and $\mathfrak{F}^{-1}[1/(p - i\tau)] = e^{i\tau t}\,e^{-\sigma t}H(t)$. Therefore,

$$\left|\int_{-\infty}^{\infty} \frac{(1 - p^2)\varphi(\omega)}{p - i\tau}\,d\omega\right| = 2\pi\left|\int_{-\infty}^{\infty} \hat{\varphi}_\sigma(t)\,e^{i\tau t}\,e^{-\sigma t}H(t)\,dt\right|$$

$$\leq 2\pi\int_0^{\infty} |\hat{\varphi}_\sigma|\,e^{-\sigma t}\,dt \leq C$$

for all $\sigma \geq 0$. Then the dominated convergence theorem allows one to conclude that

$$\lim_{\sigma \to 0} \int_{-\infty}^{\infty} dM(\tau) \int_{-\infty}^{\infty} \frac{(1 - p^2)\varphi(\omega)}{p - i\tau} d\omega$$

$$= \int_{-\infty}^{\infty} dM \lim_{\sigma \to 0} \int_{-\infty}^{\infty} \frac{(1 - p^2)\varphi(\omega)}{p - i\tau} d\omega.$$

But with $\varphi \in S$ one may obtain classically,

$$\lim_{\sigma \to 0} \int_{-\infty}^{\infty} \frac{(1 - p^2)\varphi(\omega)}{p - i\tau} d\omega$$

$$= (1 + \tau^2)\pi\varphi(\tau) + \text{pv} \int_{-\infty}^{\infty} \frac{(1 + \omega^2)\varphi(\omega)}{i(\omega - \tau)} d\omega$$

$$= (1 + \tau^2)\pi\varphi(\tau) + \text{pv} \int_{-\infty}^{\infty} \frac{[1 + (\omega + \tau)^2]\varphi(\omega + \tau)}{i\omega} d\omega,$$

or finally,

$$\lim_{\sigma \to 0} \left\langle (1 - p^2) \int_{-\infty}^{\infty} \frac{dM}{p - i\tau}, \varphi \right\rangle$$

$$= \int_{-\infty}^{\infty} (1 + \tau^2)\pi\varphi(\tau) \, dM$$

$$+ \int_{-\infty}^{\infty} dM \, \text{pv} \int_{-\infty}^{\infty} \frac{[1 + (\omega + \tau)^2]\varphi(\omega + \tau)}{i\omega} d\tau$$

$$= \langle \pi(1 + \omega^2) \, DM, \varphi \rangle + \left\langle (1 + \omega^2)\left(DM * \text{pv} \frac{1}{i\omega} \right), \varphi \right\rangle,$$

which finally gives

$$(1 - p^2) \int_{-\infty}^{\infty} \frac{dM}{p - i\tau} \xrightarrow{S'} \pi(1 + \omega^2) \, DM + (1 + \omega^2)\left(DM * \text{pv} \frac{1}{i\omega} \right).$$

The sufficiency of the theorem follows from the remark that $\overline{W}^T + W = 2\pi(1 + \omega^2)\,DM$ because this allows one to identify M, given W_ω, and then by inspection one determines Q, and A. The conditions of the theorem guarantee that the Q, A, and M so identified will satisfy Theorem 3.16. Thus the $W(p)$ formed from these parameters, by means of Youla's representation, will be positive-real. The above demonstration proves that this $W(p)$ will converge in the S' topology to the original distribution.

The theorem takes on a particularly simple form when the positive-real matrix reduces to a scalar function, i.e., the boundary value of a positive real function must be of the form $W_\omega = i\{\omega A + \omega \int_{-\infty}^{\infty} dM - (1 + \omega^2)[DM * \mathrm{pv}\,(1/\omega)]\} + \pi(1 + \omega^2)\,DM$ with $\mathrm{Re}\,W_\omega = \pi(1 + \omega^2)\,DM$, A a real positive constant, and M a bounded, odd, nondecreasing, real function. Thus DM is a nonnegative measure and we see most simply how the positiveness of the real part of the holomorphic function, $W(p)$, is reflected into the behavior of the boundary distribution. In addition the statement of the theorem can be interpreted as a specialized Hilbert transform relationship for positive real functions, i.e., the imaginary part (at least in the scalar case) of the boundary distribution is determined, with the exception of the constant A, from the real part.[9] In any event, the theorem characterizes the positive real class completely and does so by considering only the boundary values W_ω. Other concise characterizations have been obtained, notably one by König and Zemanian [K1]. In their formulation, direct use is made of the generalized Green's function. Specifically, they proved that $w(t)$ is the generalized Green's function of a passive immittance operator if and only if $w(t) = A\delta^{(1)} + w_0(t)$, where A is a real, symmetric, nonnegative definite matrix, and $w_0 \in \mathscr{D}'_+$ is a matrix of real distributions each of whose elements are second-order, weak derivatives of continuous functions such that $\bar{\mathbf{y}}^T[w_0 + \tilde{w}_0^T]\mathbf{y}$ is a nonnegative definite distribution for all constant $n \times 1$ vectors $\mathbf{y}$.

Another formulation of passive operator theory has been of considerable interest in the area of scattering theory. In this development, one defines $\mathbf{a} = \frac{1}{2}(\mathbf{u} + \mathbf{v})$, $\mathbf{b} = \frac{1}{2}(\mathbf{u} - \mathbf{v})$ and then assumes that there exists a linear, continuous, translation invariant causal operator, called the scattering operator, that maps **a** into **b**. Such an operator can be considered passive in the sense of the following.

DEFINITION 3.4.   A passive scattering operator is a linear, continuous, translation invariant, causal mapping of real $\mathbf{a} \in \mathscr{E}'$ into real $\mathbf{b} \in \mathscr{D}'$ with the property that

$$\int_{-\infty}^{t} [\bar{\mathbf{a}}^T \mathbf{a} - \bar{\mathbf{b}}^T \mathbf{b}] \, dx \geq 0 \qquad (3.24)$$

for all $\mathbf{a} \in C_0^\infty$ and all t.

This definition of passivity is not distinct from the previous one since $\int_{-\infty}^{t} [\bar{\mathbf{a}}^T \mathbf{a} - \bar{\mathbf{b}}^T \mathbf{b}] \, dx = \frac{1}{2} \int_{-\infty}^{t} [\bar{\mathbf{u}}^T \mathbf{v} + \bar{\mathbf{v}}^T \mathbf{u}] \, dx$ and one concludes that Definition 3.4 again yields a measure of energy transfer when the definition is applied to an electrical network.[10] The following theorem characterizes such operators.

Theorem 3.18.   Let s_t be the generalized Green's function of a passive scattering operator (Definition 3.4), then its Laplace transform $S(p)$ is such that in $\operatorname{Re} p > 0$
   (1) $S(p)$ is holomorphic,
   (2) $S(\bar{p}) = \overline{S(p)}$,
   (3) $1_n - \overline{S(p)}^T S(p)$ is a nonnegative definite matrix.
Such a matrix, $S(p)$, is called bounded-real.

PROOF.   Let $\mathbf{a} = \mathbf{y}\varphi$ with $\mathbf{y}$ a constant $n \times 1$ vector, φ real and $\varphi \in C_0^\infty$. But since $0 \leq \int_{-\infty}^{\infty} \bar{\mathbf{a}}^T \mathbf{a} \, dt < \infty$ for all such $\mathbf{a}$, the passivity of the operator implies that $\int_{-\infty}^{\infty} \bar{\mathbf{b}}^T \mathbf{b} \, dt = \int_{-\infty}^{\infty} \bar{\mathbf{y}}^T (s^T * \varphi)(s * \varphi)\mathbf{y} \, dt < \infty$. However, this must hold for all constant $\mathbf{y}$ vectors and if we, for example, set each element of $\mathbf{y}$ equal to zero except y_r we obtain $y_r \bar{y}_r \int_{-\infty}^{\infty} \sum_{j=1}^{n} (s_{jr} * \varphi)^2 \, dt < \infty$, or every $s_{jk} * \varphi$ must be $L_2(-\infty, \infty)$ and by Theorem 1.32, each s_{jk} must be contained in $\mathscr{D}'_{L_2}$. Now employing the arguments of Lemma 3.3 we conclude that, (with $s_{jk} \in \mathscr{D}'_{L_2}$ established above) $\int_{-\infty}^{t} [\bar{\mathbf{a}}^T \mathbf{a} - \bar{\mathbf{b}}^T \mathbf{b}] \, dx \geq 0$ for all t and all $\mathbf{a} \in S$. Then with $\mathbf{a} = \mathbf{y} \, e^{pt}$ we obtain, as in Theorem 3.15,

$$\int_{-\infty}^{t_0} [\bar{\mathbf{a}}^T \mathbf{a} - \bar{\mathbf{b}}^T \mathbf{b}] \, dx = \bar{\mathbf{y}}^T [1_n - \overline{S(p)}^T S(p)] \mathbf{y} \int_{-\infty}^{t_0} e^{2 \operatorname{Re} pt} \, dt \geq 0$$

for $\operatorname{Re} p > 0$. The other two conditions follow from the causality of the operator and the fact that $s(t)$ is real. The converse to this theorem may be established on a classic basis following the discussion of Youla, Carlin, and Castriota [Yo1].

Again, as with the immittance formulation, the distributional boundary behavior of the bounded-real matrix $S(p)$ allows us to obtain a concise representation for the passive scattering operator. Before demonstrating this result, we will obtain a theorem of independent interest in the theory of distributional boundary behavior which bears directly on this discussion.

Theorem 3.19. A necessary and sufficient condition that a bounded measurable function $\hat{u}(\omega)$ be the boundary value in the S' topology (as well as pointwise a.e.) of a bounded holomorphic function in $\operatorname{Re} p > 0$ is that

$$D\hat{u}(\omega) = \frac{1}{\pi i}\hat{u} * D \operatorname{pv} \frac{1}{\omega}. \tag{3.25}$$

PROOF. Suppose that $\hat{u}(\omega)$ is the limit of some $\hat{u}(p)$, holomorphic and bounded in $\operatorname{Re} p > 0$. Then $\hat{u}_0(p) = [\hat{u}(p)/(1 + p)] \in H^2$ and by Theorem 3.12 $\hat{u}_0(p)$ is the Laplace transform of some $u_0 \in L_2(0, \infty)$. But then $\hat{u}(p)$ is the Laplace transform of $u = Du_0 + u_0$ where $\mathfrak{F}u = \hat{u}$. The necessity of the theorem then follows from Theorem 3.10 with $k = 1$.

Conversely, suppose $\hat{u}(\omega)$ satisfies $D\hat{u} = (1/\pi i)\hat{u} * D \operatorname{pv} (1/\omega)$. Then again, from Theorem 3.10, $\hat{u}$ is the Fourier transform of some $u \in \mathscr{D}'_+$ and, since $\hat{u}_0(\omega) = [\hat{u}(\omega)/(1 + i\omega)] \in L_2, u = Du_0 + u_0$ where $\mathfrak{F}u_0 = \hat{u}_0$. The Poisson representation then shows that $\hat{u}_0(p)$, the Laplace transform of u_0, is holomorphic and bounded in $\operatorname{Re} p > 0$. To complete the proof, it then suffices to show that $p\hat{u}_0$ is bounded. To this end, we use a Phragmen–Lindelof argument. Let $G(\gamma, p) = \exp[-\gamma p^\alpha] p\hat{u}_0(p)$, $0 < \alpha < 1$. Moreover, let r, θ define $p, |\theta| < \pi/2$. Then $G(\gamma, p)$ is holomorphic for $\operatorname{Re} p > 0$ and, for every fixed $\gamma > 0$,

$$|G(\gamma, p)| \leq \exp[-\gamma r^\alpha \cos \alpha\theta] \, r|\hat{u}_0(p)|$$

is uniformly bounded in Re $p > 0$. Further, as $\sigma \to 0$, $G(\gamma, p) \to G(\gamma, \omega)$ $= \exp[-\gamma|\omega|^\alpha \cos \pi(\alpha/2)](i\omega)\hat{u}_0(\omega)$, a.e. and $|G(\gamma, \omega)| \leq M$. Hence $G(\gamma, p)$ is uniformly bounded in γ for $\sigma > 0$. Finally, $|p\hat{u}_0(p)| \leq |G(\gamma, p)|$ $\exp[\gamma r^\alpha \cos \alpha\theta]$ so that, if we let $\gamma \to 0$ for fixed r, θ, we conclude that $\hat{u}(p) = (1 + p)\hat{u}_0(p)$ is bounded. Theorem 3.12 shows that $\hat{u}(p)$ converges in the S' topology to $\hat{u}(\omega)$ and is unique.

This theorem is, in a sense, a converse to the classic Fatou result that states that bounded holomorphic functions converge pointwise a.e. to bounded measurable functions as $\sigma \to 0$.[11] We may obtain an interesting collateral result if we assume that a given holomorphic function converges in the $\mathscr{D}'$ topology (as we saw in Chapter II, a much weaker convergence than S') to a distribution which is a bounded measurable function satisfying the condition of Theorem 3.19. The conclusion that such a holomorphic function must be bounded in Re $p > 0$ follows from Theorem 3.19 and the extended $\mathscr{D}'$ uniqueness theorem for distributional boundary values described in Section 3.4. This extension of Theorem 3.19 constitutes a maximum modulus theorem in which the condition under which it is applicable must be interpreted distributionally even though the conclusion of the theorem is classical, i.e., implies point-wise convergence a.e. In particular, if the boundary value $\hat{u}_\omega$ satisfies $|\hat{u}_\omega| \leq M$, then from a classic representation theorem for bounded functions (see Nevanlinna [N1])

$$\hat{u}(p) = \frac{\sigma}{\pi} \int_{-\infty}^{\infty} \frac{\hat{u}(\zeta)}{(\zeta - \omega)^2 + \sigma^2} d\zeta, \text{Re } p > 0,$$

and

$$|\hat{u}(p)| = \frac{1}{\pi} \left| \sigma \int_{-\infty}^{\infty} \frac{\hat{u}(\zeta)}{(\zeta - \omega)^2 + \sigma^2} d\zeta \right| \leq M \frac{\sigma}{\pi} \int_{-\infty}^{\infty} \frac{1}{(\zeta - \omega)^2 + \sigma^2} d\zeta = M$$

in Re $p > 0$.

Returning to the problem of characterizing passive scattering operators we employ Theorem 3.19 to obtain the following theorem.

Theorem 3.20. The necessary and sufficient conditions that an $n \times n$ matrix of measurable function $S(\omega)$ be the boundary behavior point-wise a.e., as well as in the S' topology, of a bounded-real matrix are

(1) $1_n - \overline{S(\omega)}^T S(\omega)$ be nonnegative definite a.e.,

(2) $DS(\omega) = (1/\pi i)S * D$ pv $(1/\omega)$, and

(3) $S(-\omega) = \overline{S(\omega)}$.

PROOF. The necessity of the third condition follows from the fact that $S(\omega) = \mathfrak{F}s$ where s is a real distribution. Moreover, from Theorem 3.18, the fact that $1_n - \overline{S(p)}^T S(p)$ is nonnegative definite in Re $p > 0$ implies that each element of $S(p)$ is bounded in Re $p > 0$. Condition (2) then follows from Theorem 3.19 since each element of $S(p)$ is known to be holomorphic in Re $p > 0$. Finally, the bounded nature of each element of $S(p)$ implies, via the Fatou theorem, that they separately converge point-wise a.e. to bounded measurable functions on the boundary. Therefore,

$$\lim_{\sigma \to 0} \bar{\mathbf{y}}^T [1_n - \overline{S(p)}^T S(p)]\mathbf{y} = \bar{\mathbf{y}}^T [1_n - \overline{S(\omega)}^T S(\omega)]\mathbf{y}$$

exists a.e. and since $\bar{\mathbf{y}}^T [1_n - \overline{S(p)}^T S(p)]\mathbf{y} \geq 0$ for all Re $p > 0$, the limit must also be nonnegative, i.e., condition (1) is established.

As to the sufficiency of these statements; (1) and (2) imply on the basis of Theorem 3.19 that $S(p) = \mathcal{L}[\mathfrak{F}^{-1}S(\omega)]$ is a matrix of bounded holomorphic functions in Re $p > 0$ that have $S(\omega)$ as their boundary value. The maximum modulus aspect of Theorem 3.19 implies that

$$\bar{\mathbf{y}}^T [1_n - \overline{S(p)}^T S(p)]\mathbf{y} = \bar{\mathbf{y}}^T \mathbf{y} - [\overline{S(p)\mathbf{y}}]^T [S(p)\mathbf{y}]$$

$$= \sum_{k=1}^{n} |y_k|^2 - \sum_{j=1}^{n} \left| \sum_{k=1}^{n} S_{jk}(p)y_k \right|^2$$

$$\geq \bar{\mathbf{y}}^T \mathbf{y} - [\overline{S(\omega)\mathbf{y}}]^T [S(\omega)\mathbf{y}] \geq 0$$

or $1_n - \overline{S(p)}^T S(p)$ is nonnegative definite in Re $p > 0$. Finally, since $S(-i\omega) = \overline{S(i\omega)}$, $\mathfrak{F}^{-1}S(\omega)$ is real and so $\mathcal{L}(\mathfrak{F}^{-1}S) = S(p)$ satisfies $S(\bar{p}) = \overline{S(p)}$.

We have completed our program of studying passive operators to the extent that Theorems 3.17 and 3.20 offer a concise representation of these operators.[12] There is, however, considerable interest in a class of passive operators, particularly in the theory of electromagnetic waves, for which the passivity (or energy) condition takes on the value zero for $t \to \infty$. Such operators are referred to as unitary or lossless operators and we specifically define them by

DEFINITION 3.5.   A passive scattering operator is conservative or lossless if

$$\int_{-\infty}^{\infty} [\bar{\mathbf{a}}^T\mathbf{a} - \bar{\mathbf{b}}^T\mathbf{b}] \, dt = 0 \tag{3.26}$$

for all $\mathbf{a} \in C_0^\infty$.

The characterization of such operators can be obtained as a special case of the result concerning passive operators. In fact, the Laplace transform of the generalized Green's function of the operator must satisfy Theorem 3.18, i.e., it must be bounded real. However, the precise characterization of these operators can only be stated on the boundary. Thus we have the following theorem.

Theorem 3.21.   The necessary and sufficient condition that an $n \times n$ bounded-real matrix represent a lossless or unitary operator is that its pointwise, a.e., boundary value matrix $S(\omega)$ consist of measurable functions and $1_n = \overline{S(\omega)}^T S(\omega)$, a.e.

PROOF.   The sufficiency of the condition is established by employing the Parseval relation to obtain, for $\mathbf{a} \in C_0^\infty$,

$$\int_{-\infty}^{\infty} (\bar{\mathbf{a}}^T\mathbf{a} - \bar{\mathbf{b}}^T\mathbf{b}) \, dt = \frac{1}{2\pi} \int_{-\infty}^{\infty} \bar{\boldsymbol{\psi}}^T [1_n - \overline{S(\omega)}^T S(\omega)] \boldsymbol{\psi} \, d\omega = 0,$$

where $\boldsymbol{\psi} = \mathfrak{F}\mathbf{a}$.

The necessity of the condition may be established by noting that for $\mathbf{a} = \mathbf{y}\varphi$,

$$\int_{-\infty}^{\infty} [\bar{\mathbf{a}}^T\mathbf{a} - \bar{\mathbf{b}}^T\mathbf{b}] \, dt = \int_{-\infty}^{\infty} (\bar{\mathbf{y}}^T\bar{\varphi}\mathbf{y}\varphi - [\bar{\mathbf{y}}^T s^T * \bar{\varphi}][s * \varphi]\mathbf{y}) \, dt$$

$$= \langle \bar{\mathbf{y}}^T[\delta 1_n - \tilde{s}^T * s]\mathbf{y}, \bar{\varphi} * \tilde{\varphi} \rangle = 0,$$

which follows using the arguments developed in the proof of Lemma 3.2. However, as we saw in the proof of Lemma 3.2, $\langle \chi, \varphi * \tilde{\bar{\varphi}} \rangle = 0$ certainly implies $\chi * \varphi * \tilde{\bar{\varphi}}$ is a continuous nonnegative definite function which, as is known classically, is bounded in magnitude by its value at the origin. But $\langle \chi, \varphi * \tilde{\varphi} \rangle = \chi * \varphi * \tilde{\varphi}|_{t=0} = 0$ or $\chi * \varphi * \tilde{\varphi} = 0$ for all t, which implies, since φ is an arbitrary member of C_0^∞, that $\chi = 0$. Therefore we obtain $\bar{\mathbf{y}}^T[\delta 1_n - \tilde{s}^T * s]\mathbf{y} = 0$, and taking Fourier transforms yields $\bar{\mathbf{y}}^T[1_n - \overline{S(\omega)}^T S(\omega)]\mathbf{y} = 0$ a.e.

But since $\mathbf{y}$ is arbitrary, we conclude that

$$1_n = \overline{S(\omega)}^T S(\omega) \quad \text{a.e.} \tag{3.27}$$

3.6 Notes and Remarks

1. The notion of Cauchy integral of a distribution was exploited by Bremermann and Durand [Br3] where, in particular, a proof of necessity of the conditions in Theorem 3.3, is to be found. It seems worthwhile to review briefly here an interesting integral representation for distributions in $\mathscr{E}'$ [Br3]. Let $\mathscr{A}$ denote the class of testing functions which are holomorphic in some fixed nontrivial strip containing the ω axis. Clearly, $\mathscr{A} \subset \mathscr{E}$. If f_ω is a distribution of compact support K then, for any $\varphi \in \mathscr{A}$,

$$\langle f, \varphi \rangle = \int_\Gamma C(f, p)\varphi(p)\, dp, \tag{3.28}$$

where Γ is any closed contour taken around K and contained within the strip of analyticity. This representation is independent of the Γ chosen.

The problem of how to multiply distributions is crucial for renormalization problems in quantum field theory (see Bremermann [Br4]). By forming finite parts of expressions of type (3.28) it appears that a reasonable multiplication theory is possible without the usual divergence difficulties. In this regard see also the AMS abstract by Tillman [Ti1].

Most of the results in this section on Cauchy integrals are new and unpublished outgrowths of the authors' earlier work (Beltrami and Wohlers [B2]).

2. The Hilbert and Wiener-Hopf problems are discussed in depth in the books by Muskhelishvili [M1] and Noble [No1], respectively. One should note that in order to complete the program of studying Wiener-Hopf factorizations distributionally one must apparently give meaning to the logarithm of a distribution. Unfortunately, however, no effective functional calculus for distributions is available as yet to do this.

3. Equations (3.12) and (3.13) of this corollary extend the Plemelj formulas for the boundary values of half plane Cauchy integrals (in the classical situation $k = 0$). The Plemelj relations are discussed in detail in Muskhelishvili [M1] for the case in which the boundary functions are Hölder continuous.

4. Theorem 3.10 was proven by Beltrami and Wohlers [B3]. The standard reference for the corresponding classical result is Titchmarsh [T1], Chapter 5.

5. The relation between causality and dispersion has been studied by many authors all following, in a sense, the L_2 work of Titchmarsh, referenced above. In this regard it suffices to mention the papers by Bremermann, Oehme, and Taylor [Br2], Toll [To1], and Taylor [Ta1].

6. This continuation theorem can be extended to functions of n complex variables and to $\mathscr{D}'$ distributions (see Appendix II) as the "Edge of the Wedge" theorem. Our proof for S' follows an idea to be found in Taylor [Ta1]. The uniqueness Corollary 3.4 was established by different arguments by Beltrami and Wohlers [B3].

7. The important classes of H^+ functions to be studied in this section concern those functions $f(p)$ which either map the half plane into itself or the half plane into the disk [actually, we make an additional symmetry requirement that $f(p)$ is real for p real: as we will see this is equivalent to requiring the operator kernel to be real].

8. The notion of dissipative operator also appears in a Hilbert space setting, as given by Phillips (see the review article by Dolph [D1]). The resolvent $(pI - A)^{-1}$ of a closed dissipative operator A has analyticity and representation properties analogous to the ones studied here except that semigroup methods are used to establish

them. It is worth noting that the resolvent or immittance need not be rational, and may correspond to distributed systems.

An important observation of Youla, Castriota, and Carlin [Yo1] is that passivity, in the sense of Definition 3.3, implies causality for linear operators. The following Theorem 3.15 could then be stated as: a linear, continuous, translation invariant, and passive operator has a positive real resolvent.

9. The scalar Cauer formula can be given in alternate forms. Thus if we let $[(1 - pi\tau)/(p - i\tau)] = -i\tau + [(1 + \tau^2)/(p - i\tau)]$, and note that $M(\tau)$ is odd, then

$$
\begin{aligned}
f(p) &= Ap + \int_{-\infty}^{\infty} \frac{(1 + \tau^2)\,dM(\tau)}{p - i\tau} \\
&= Ap + p \int_{-\infty}^{\infty} \frac{(1 + \tau^2)\,dM(\tau)}{p^2 + \tau^2}.
\end{aligned}
\tag{3.29}
$$

Now pass to the limit in (3.29), in the S' topology, as $\sigma \to 0$, to obtain

$$
\begin{aligned}
\operatorname{Re} f(p) &\to \pi(1 + \omega^2)\,DM_\omega \\
\operatorname{Im} f(p) &\to A\omega - \frac{1}{\pi}\operatorname{Re} f(i\omega) * \operatorname{pv} 1/\omega,
\end{aligned}
\tag{3.30}
$$

which closely resembles a Hilbert transform relation (actually, it can be shown that precise Hilbert pairs (3.17) and (3.18) hold here with $k = 2$). Equation (3.30) has to be interpreted in a distributional sense and is not attainable classically. Thus if $f(p) = 1/p$ then (3.30) shows that $\operatorname{Re} f(i\omega) = \pi\delta$ and $\operatorname{Im} f(i\omega) \equiv \operatorname{pv} 1/\omega$. The best that can be hoped for (classically) are partial statements on the boundary. For example, if $M(\tau)$ is absolutely continuous then one can obtain from (3.29) that

$$
\operatorname{Im} f(i\omega) = A\omega - 2\omega\operatorname{pv} \int_0^{\infty} \frac{(1 + \tau^2)\,dM(\tau)}{\omega^2 - \tau^2} \qquad \text{(see Youla [Yo2]).}
$$

If M is not absolutely continuous this relation holds only a.e., and for $1/p$ this means that $\operatorname{Im} f(i\omega) = -1/\omega$ a.e., which is an incomplete statement (here $A = 0$). Under restrictive assumptions similar formulas occur classically in dispersion theory as the Kramers-Kronig relations (see, e.g., Toll [To1]).

10. It has been shown by Wohlers and Beltrami [Wo1] that if passivity is defined as in Definition 3.4 but, however, with the upper limit of the integral equal to ∞, then a complete theory (equivalent to the one described in this chapter) can be obtained only when causality is invoked as an additional postulate.

11. Theorem 3.19 was first presented by the authors at a meeting of the AMS [B4] and published by Beltrami and Wohlers [B3]. The fact that $f(p) \in H^\infty$ has a limit a.e. as $\sigma \to 0$ was established by Fatou [Fa1].

12. Youla, Castriota, and Carlin [Yo1] completely characterized passive scattering operators when they map L_2 into itself. Wohlers and Beltrami [Wo1] found the operator kernel to be a distribution in $\mathscr{D}'_{L_2}$; the boundary description (frequency behavior) of such operators (Theorem 3.20) is also established there. In the case of passive immittance operators the basic characterization of the Green's function (time domain behavior) is due to Zemanian [Z2] (Theorem 3.15 and its converse), a result which is refined in König and Zemanian [K1]. A complete description of such operators in terms of the frequency behavior is given by Theorem 3.17, and was established (in the scalar case) by Wohlers and Beltrami [Wo1]. The matrix versions of Theorems 3.17 and 3.20 were given by Wohlers [Wo2]. Note that in the matrix case the real part of the boundary values is given by the hermetian real part $\overline{W}_\omega^T + W_\omega$ since the real part of the matrix itself yields no particular information.

Finally, it should be noted here that linear operators arising in network theory have also been studied by Newcomb [Ne1], from a distributional point of view.

APPENDIX I

Representation of Positive-Real Matrices

We will develop the material necessary to prove Youla's representation theorem for positive-real matrices (Theorem 3.16) by returning to a classic result of Herglotz.

Theorem A1.1. A function $h(\xi)$ is analytic and has a positive real part in the circle $|\xi| < 1$ if and only if

$$h(\xi) = \frac{1}{2\pi} \int_0^{2\pi} \frac{e^{i\alpha} + \xi}{e^{i\alpha} - \xi} \, d\psi(\alpha) + iq,$$

where $\psi(\alpha)$ is a real nondecreasing function of bounded variation on $[0, 2\pi)$, and q is a real constant.

For a proof of this theorem we refer the reader to Nevanlinna [N1].

For our purposes, we require a representation for functions with positive real parts in a half plane. The following corollary (Cauer [C1]) is the desired statement.

Corollary A1.1. A function $f(p)$ is analytic and has a positive real part in the half plane $\operatorname{Re} p > 0$ if and only if

$$f(p) = \int_{-\infty}^{\infty} \frac{1 - pi\tau}{p - i\tau} \, d\beta(\tau) + ap + iq,$$

99

where $\beta(\tau)$ is a real nondecreasing function of bounded variation on $(-\infty, \infty)$ a a nonnegative real constant, and q a real constant. If in addition, $f(\bar{p}) = \overline{f(p)}$, i.e., $f(p)$ is a positive-real function, then we may select $\beta(\tau) = -\beta(-\tau)$ and have $q = 0$.

PROOF.   The transformation

$$\xi = \frac{p-1}{p+1} \quad \text{or} \quad p = \frac{1+\xi}{1-\xi}$$

maps the interior of the unit circle in the "ξ" plane into the right half of the "p" plane in a conformal manner. In particular, the circumference of the circle maps into the real axis via the transformation

$$e^{i\alpha} = \frac{i\tau - 1}{i\tau + 1} \quad \text{or} \quad i\tau = \frac{1 + e^{i\alpha}}{1 - e^{i\alpha}}.$$

Now if we let $-(1/2\pi)\psi(\alpha) = \beta(\tau)$, the integral of Theorem A1.1 may be expressed, via a change of variables, as

$$h(\xi) = f(p) = \int_{-\operatorname{ctn} \varepsilon/2}^{\operatorname{ctn} \varepsilon/2} \frac{1 - pi\tau}{p - i\tau} d\beta(\tau) + \frac{1}{2\pi} \int_0^\varepsilon \frac{e^{i\alpha} + \xi}{e^{i\alpha} - \xi} d\psi(\alpha)$$

$$+ \frac{1}{2\pi} \int_{2\pi-\varepsilon}^{2\pi} \frac{e^{i\alpha} + \xi}{e^{i\alpha} - \xi} d\psi(\alpha) + iq, |\xi| < 1, \operatorname{Re} p > 0$$

for some positive ε. Note that we have mapped the portion of the circumference $\varepsilon < \theta < 2\pi - \varepsilon$ into the corresponding segment of the real axis in the "p" plane. Letting $\varepsilon \to 0$ we obtain

$$f(p) = \int_{-\infty}^{\infty} \frac{1 - pi\tau}{p - i\tau} d\beta(\tau) + \frac{1+\xi}{1-\xi} \left[\frac{\psi(0^+) - \psi(0)}{2\pi} \right]$$

$$+ \frac{1+\xi}{1-\xi} \left[\frac{\psi(2\pi) - \psi(2\pi^-)}{2\pi} \right] + iq.$$

If we let $a = (1/2\pi)[\psi(0^+) - \psi(0) + \psi(2\pi) - \psi(2\pi^-)]$ we have the stated result. The symmetry statement, when $f(\bar{p}) = \overline{f(p)}$, follows directly from the basic representation as we may establish by substitution.

We are now in a position to prove Theorem 3.16 in essentially the same way as Youla presented it [Yo2].

Theorem 3.16.   The necessary and sufficient condition that an $n \times n$ matrix $W(p)$ be positive-real is that for $\text{Re}\, p > 0$,

$$W(p) = Q + Ap + \int_{-\infty}^{\infty} \frac{1 - pi\tau}{p - i\tau} \, dM(\tau),$$

where Q is a real, constant, skew-symmetric $n \times n$ matrix ($Q^T = -Q$); A is a real, constant, symmetric, nonnegative definite matrix; and M is an $n \times n$ matrix each of whose elements are of bounded variation on the entire line $-\infty < \tau < \infty$, satisfying $\overline{M}^T = M$, $-\overline{M}(-\tau) = M$, and with $\overline{\mathbf{a}}^T M(\tau)\mathbf{a}$ being a real, bounded nondecreasing function of τ for any complex constant vector $\mathbf{a}$.

PROOF.   Since $W(p)$ is positive real

$$\overline{\mathbf{y}}^T[W(p) + \overline{W(p)}^T]\mathbf{y} \equiv 2\,\text{Re}[\overline{\mathbf{y}}^T W(p)\mathbf{y}] \geq 0$$

in $\text{Re}\, p > 0$, and for all complex constant vectors $\mathbf{y}$. In particular, if every element of $\mathbf{y}$ is zero except y_r and y_k then

$$\text{Re}\{\overline{y}_r y_r W_{rr}(p) + \overline{y}_r y_k W_{rk}(p) + y_r \overline{y}_k W_{kr}(p) + \overline{y}_k y_k W_{kk}(p)\} \geq 0.$$

Thus the function $a_1(p) = W_{rr}(p) + W_{rk}(p) + W_{kr}(p) + W_{kk}(p)$ has a positive real part, as we note when $y_r = y_k = 1$. However, since each element satisfies $\chi(\overline{p}) = \overline{\chi(p)}$ and in addition is holomorphic in $\text{Re}\, p > 0$ we conclude that $a_1(p)$ is itself a positive-real function. Similarly, if we let $y_r = -y_k = 1$ we observe that $a_2(p) = W_{rr}(p) -[W_{rr}(p) + W_{kr}(p)] + W_{kk}(p)$ is also a positive-real function. In addition, if we let $y_r = i, y_k = 1$ and then $y_r = i, y_k = -1$, we observe that

$$b_1(p) = W_{rr}(p) - i[W_{rk}(p) - W_{kr}(p)] + W_{kk}(p)$$

and $b_2(p) = W_{rr}(p) + i[W_{rk}(p) - W_{kr}(p)] + W_{kk}(p)$ have positive real parts and are holomorphic in $\text{Re}\, p > 0$ (they are not positive-real since they fail the symmetry property). Therefore, $W_{rk}(p) + W_{kr}(p) = \frac{1}{2}[a_1(p) - a_2(p)]$ is the difference between two positive real functions, and $i[W_{rk}(p) - W_{kr}(p)] = \frac{1}{2}[b_2(p) - b_1(p)]$ is the difference between two analytic functions with positive real parts. Employing Corollary A1.1 we obtain the representation

$$W_{rk}(p) + W_{kr}(p) = ap + \int_{-\infty}^{\infty} \frac{1 - pi\tau}{p - i\tau} \, d\beta(\tau),$$

where a is real (not necessarily positive), and $\beta(\tau) = -\beta(-\tau)$ is a bounded function (not necessarily nondecreasing). In addition,

$$W_{rk}(p) - W_{kr}(p) = q - iap - i \int_{-\infty}^{\infty} \frac{1 - pi\tau}{p - i\tau} \, d\theta(\tau),$$

where q and a are real and $\theta(\tau)$ is bounded. However, since $W_{rk}(\bar{p}) - W_{kr}(\bar{p}) = \overline{W_{rk}(p)} - \overline{W_{kr}(p)}$ we observe that $a = 0$ and $\theta(-\tau) = \theta(\tau)$, or

$$W_{rk}(p) - W_{kr}(p) = q - i \int_{-\infty}^{\infty} \frac{1 - pi\tau}{p - i\tau} \, d\theta(\tau).$$

Thus we may conclude, considering each element of the matrix $W(p)$ separately, that if $W(p)$ is positive-real

$$W(p) + W^{T}(p) = Ap + \int_{-\infty}^{\infty} \frac{1 - pi\tau}{p - i\tau} \, d\beta(\tau)$$

$$W(p) - W^{T}(p) = Q + \int_{-\infty}^{\infty} \frac{1 - pi\tau}{p - i\tau} \, d[-i\theta(\tau)],$$

where A is a real, symmetric $n \times n$ matrix of constants, Q is a real skew-symmetric $(Q^{T} = -Q)$ $n \times n$ matrix of constants, $\beta(\tau)$ is a symmetric $n \times n$ matrix of odd bounded functions, and $\theta(\tau)$ is a skew-symmetric $n \times n$ matrix of even bounded functions. By addition $\beta(\tau) - i\theta(\tau) = M(\tau)$ is an $n \times n$ matrix of bounded functions satisfying $\overline{M}^{T} = M$ and $-\overline{M}(-\tau) = M$. The theorem follows finally by adding the two matrix expressions and noting that

$$\bar{\mathbf{y}}^{T}[\overline{W}^{T} + W]\mathbf{y} = 2[\bar{\mathbf{y}}^{T}A\mathbf{y}]p + \int_{-\infty}^{\infty} \frac{1 - pi\tau}{p - i\tau} \, d\{\bar{\mathbf{y}}^{T}[\overline{M}^{T} + M]\mathbf{y}\} \geq 0$$

in Re $p > 0$ implies via Corollary A1.1 that $\bar{\mathbf{y}}^T A \mathbf{y}$ must be nonnegative, i.e., A is nonnegative definite, and that $\bar{\mathbf{y}}^T [\overline{M}^T + M] \mathbf{y} = 2\bar{\mathbf{y}}^T M(\tau)\mathbf{y}$ must be a real, bounded, nondecreasing function of τ for all complex vectors $\mathbf{y}$.

APPENDIX II

Supplementary Remarks

Introduction

It is safe to say that up to now the most striking and successful application of the theory of distributions have been to questions in the theory of linear partial differential operators and in the study of the boundary behavior of analytic functions of one and several complex variables. In Chapter III we took a close look at the boundary value question for functions of a single variable, in connection with the Laplace transform of those causal and tempered distributions which describe the Green's function of certain linear and time invariant operators. When such operators are defined by means of differential expressions on E^n, certain explicit existence questions can be resolved as we will briefly indicate in this Appendix. Moreover, it will be of interest to look at how the question of the boundary behavior of functions of several complex variables leads to a natural extension of the ideas pursued in Chapter III. It is in this extended setting that many of the problems of quantum field theory find their appropriate framework. The authors' own work lies in the simpler setting of Chapter III and so, in this survey, we make no pretense at completeness.

A.1 PARTIAL DIFFERENTIAL OPERATORS

In Section 3.1 we considered a class of linear operators which map $\mathscr{E}'$ into $\mathscr{D}'$. In systems terminology such an operator T defines a black box for which given inputs $f \in \mathscr{E}'$ are linearly transformed into outputs Tf. If the system is described by a partial differential operator P with constant coefficients (see precise definition below), then $T = P^{-1}$ in the sense that whenever f is a given input in $\mathscr{E}'$ we can find a unique output $u \in \mathscr{D}'$, which satisfies $Pu = f$, and so $u = P^{-1}f \equiv Tf$. In this section we want to outline a proof of the fact that a solution $u \in \mathscr{D}'$ always exists for the equation $Pu = f, f \in \mathscr{E}'$. Moreover, such a u is given by $f * v$ where v satisfies $Pv = \delta$; now the linearity, translation invariance, and continuity of P implies the same for P^{-1}, and so the fact that $v = T\delta$ is not surprising, in view of Theorem 1.18.

In this appendix we work on E^n and so the following definitions are needed.

Let $l = (l_1, \ldots, l_n)$, with l_i nonnegative integers, and let $|l| = \Sigma_{i \leq n} l_i$. If $D_i = (\partial/\partial x_i)$ then $D^l = D_1^{l_1} D_2^{l_2} \ldots D_n^{l_n}$. The notation $|x|$ for any $x = (x_1, \ldots, x_n) \in E^n$ is used to designate the norm $(\Sigma_{i \leq n} x_i^2)^{1/2}$.

Now let $P(\zeta) = \Sigma_{|l| \leq m} a_l \zeta^l$ be a polynomial on E^n of degree m (a_l constants). This polynomial allows us to define a partial differential operator of constant coefficients and order m by

$$P(D) = \sum_{|l| \leq m} a_l D^l. \tag{A2.1}$$

As shown in the Introduction to Chapter I, a distribution u is defined to be the solution of $P(D)u = f$ (we write $Pu = f$), whenever

$$\langle f, \varphi \rangle = \langle u, \check{P}\varphi \rangle \tag{A2.2}$$

for all $\varphi \in C_0^\infty$; here $\check{P}$ is the operator $P(-D)$ and C_0^∞ will consist of all infinitely differentiable functions on E^n having compact support. $\mathscr{D}$ denotes C_0^∞ with a suitable topology which mimics that of Section 1.1.

A distribution v, which satisfies $Pv = \delta$, is called a fundamental solution. The existence of such a v was first established by Ehrenpreis and Malgrange (see [Ho1]), and one proof depends on the fact that for every $\alpha > 0$ and every polynomial $P(\zeta)$ one can find a constant C_1 such that

$$\|\varphi\|_{0,2} \leq C_1 \|e^{\alpha|x|} \check{P}\varphi\|_{0,2} \qquad \text{(A2.3)}$$

for all $\varphi \in C_0^\infty$, and where $\|\quad\|_{0,2}$ is the L_2 norm on E^n. Although this L_2 estimate will not be proven here it is of interest to sketch through the rest of the argument. Now in Lemma 1.6 (where $n = 1$) we showed that $|\varphi(x)| \leq \text{constant} \|\varphi\|_{1,2}$ for all $\varphi \in C_0^\infty$. In E^n, $n \geq 1$, the same type of estimate still holds except that now it reads as

$$|\varphi(x)| \leq C_2 \|\varphi\|_{m,2} = C_2 \left(\sum_{|l| \leq m} \|D^l \varphi\|_{0,2}^2 \right)^{1/2}. \qquad \text{(A2.4)}$$

for all integers $m > n/2$, and where C_2 is some other constant. The inequality (A2.4) is a form of Sobolev's Lemma (for proof, see Yosida [Y1]). Taking $x = 0$ and combining (A2.4) with (A2.3) one has, for a constant C,

$$|\varphi(0)| \leq C \left(\sum_{|l| \leq m} \|e^{\alpha|x|} D^l \check{D}\varphi\|_{0,2}^2 \right)^{1/2}. \qquad \text{(A2.5)}$$

Now let $\psi \in C_0^\infty$; then

$$\left(\sum_{|l| \leq m} \|e^{\alpha|x|} D^l \psi\|_{0,2}^2 \right)^{1/2} \qquad \text{(A2.6)}$$

is a continuous seminorm on $\mathscr{D}$. The subspace of functions in C_0^∞, which can be written as $\check{D}\varphi$ for $\varphi \in C_0^\infty$, can be mapped linearly and continuously to the complex scalars by $\check{D}\varphi \to \varphi(0)$. We now invoke the Hahn–Banach theorem which tells us, in this case, that if a linear functional F on a subspace of $\mathscr{D}$ satisfies $|F| \leq p$, for some continuous seminorm p, then F can be continuously extended to all of $\mathscr{D}$ (the Hahn–Banach theorem is proven in the book by Yosida [Y1]): thus, by virtue of (A2.5), the mapping $\check{D}\varphi \to \varphi(0)$ can be extended in a linear and continuous way to all of $\mathscr{D}$. Hence there exists a $v \in \mathscr{D}'$ such that $\langle v, \check{D}\varphi \rangle = \varphi(0)$ when $\varphi \in C_0^\infty$. But $\langle v, \check{D}\varphi \rangle = \langle Dv, \varphi \rangle$, which shows that $Dv = \delta$. To complete the existence argument let $f \in \mathscr{E}'$ and form $f * v$ (which certainly exists); then $P(f * v) = f * Pv = f * \delta = f$, as we know from Theorem 1.19. Hence $u = f * v$ is a solution to $Pu = f$, whenever $f \in \mathcal{E}'$. Note that the fundamental solution v is not unique since we can always add to it a solution of $Pu = 0$. However, when u is itself in $\mathcal{E}'$ then, taking Fourier transforms, one has that $P(D)u = 0$ implies $P(i\omega)\hat{u} = 0$; this last product cannot vanish unless $\hat{u}$ vanishes identically, since $\hat{u}$ is an entire function (Theorem 2.8) and therefore $u = 0$ which, in this case at least, implies uniqueness. The distribution v is what we called the Green's function in Chapter III. In that setting, v was further restricted by the requirement of causality.

We should note here that if $f \in \mathcal{D}'$, then a solution $u \in \mathcal{D}'$ still exists. The method of proof is in the same spirit as that outlined above for the fundamental solution. The essential ingredient is to establish an estimate for $\langle f, \varphi \rangle$ of the type studied in Theorem 1.2. Without going into details it suffices to say here that one can establish the existence of a continuous seminorm ρ on C_0^∞ for which

$$|\langle f, \varphi \rangle| \le \rho(\check{D}\varphi) \tag{A2.7}$$

for all $\varphi \in C_0^\infty$. But then one reasons as before and the linear functional which maps the subspace of $\mathcal{D}$ consisting of functions $\check{D}\varphi$ can, by (A2.7), be extended as a continuous linear functional on $\mathcal{D}$ by using the Hahn–Banach theorem, as we did earlier. Hence there exists a $u \in \mathcal{D}'$ such that $\langle Pu, \varphi \rangle = \langle u, \check{P}\varphi \rangle = \langle f, \varphi \rangle$, which is what one wants to show.

The basic existence questions for linear partial differential operators are surveyed in the articles by Schwartz [Sc3] and Gårding [Ga1]. A comprehensive treatment is to be found in the books by Hormander [Ho1] and Friedman [F1]. Extensive bibliographical information is available in these references.

A.2 FUNCTIONS OF SEVERAL COMPLEX VARIABLES AND SLOW GROWTH DISTRIBUTIONS ON E^n

Let C^n denote n-vectors $p = (p_1, \ldots, p_n)$ where $p_j = \sigma_j + i\omega_j$ are complex numbers. A function $f(p) \equiv f(p_1, \ldots, p_n)$ is holomorphic on an open connected set $\Omega \subset C^n$ (a domain in C^n) if, for every $\zeta \in \Omega$, there exists a multiple power series

$$\sum_{l_1 \cdots l_n = 0}^{\infty} a_{l_1 \cdots l_n}(p_1 - \zeta_1)^{l_1} \cdots (p_n - \zeta_n)^{l_n}, \tag{A2.8}$$

which converges uniformly to $f(p)$ in some polydisc region defined by $|p_j - \zeta_j| \le R_j$. An important result of Hartog (see [Wi1]) is that $f(p)$ is holomorphic on Ω iff it is holomorphic in each variable separately. Certain special domains will figure in our discussion. E^n is a subset of C^n, and we speak of a forward cone Γ^+ consisting of all real n-tuples $(x_1, \ldots, x_n)$ in E^n for which $x^2 = x_1^2 - \Sigma_{i=2}^n x_i^2 > 0$ and $x_i > 0$. When $x_1 < 0$ then one has a backward cone Γ^-. In one variable the forward cone is the positive axis $x_1 > 0$ (in Chapter III, x_1 was labeled t; if t refers to time then Γ^+ is the forward light cone). Corresponding to Γ^+ is a forward tube domain $T^+ \equiv \Gamma^+ + iE^n$, which consists of all complex n-tuples p for which $\operatorname{Re} p_j = \sigma_j \in \Gamma^+$. Similarly, one defines a backward tube domain T^-. When $n = 1$, T^+ is simply the half plane $\operatorname{Re} p > 0$. If Ω_1 and Ω_2 are two overlapping domains in C^n and if f_1, f_2 are holomorphic functions on Ω_1, Ω_2, respectively (which coincide on the overlap), then they are analytic continuations of each other.

We let S denote C_0^∞ functions φ on E^n for which

$$\sup_{x \in E^n} |x^m D^l \varphi| < \infty \tag{A2.9}$$

for all m (here $x^m = x_1^{m_1} x_2^{m_2} \cdots x_n^{m_n}$, $\Sigma m_i = n$). A topology on S is introduced by the seminorms (A2.9), as in Section 1.7, and the dual space S' consists of tempered or slow-growth distributions on E^n. A distribution $u \in S'$ is *causal* if its support is in the forward cone Γ^+. The Fourier transform of $\varphi \in S$ is given by

$$\tilde{\varphi}(\omega) \equiv \mathscr{F}\varphi = \int_{E^n} \varphi(x) e^{-i(\omega, x)} dx \tag{A2.10}$$

where

$$(\omega, x) = \left(\sum_{i \le n} \omega_i x_i \right)^{1/2},$$

and it can be extended as a one-one bicontinuous mapping of S' onto itself as in Section 1.8. Then the important Theorem 2.7 extends naturally to read as follows.

Theorem A2.1.   Let $u \in S'$ be causal. Then the Laplace transform $\mathfrak{F}(ue^{-(\sigma,x)})$ is holomorphic in the tube T^+ and converges in the S' sense to $\mathfrak{F}u$, as $\sigma = (\sigma_1, \ldots, \sigma_n) \to 0$. Conversely, if $f(p)$ is holomorphic in T^+ and if it has an S' limit $\hat{u}$ as $\sigma \to 0$ then $f(p) = \mathfrak{F}(ue^{-(\sigma,x)})$, and u is causal.

Certain slow-growth distributions, called vacuum expectation values, occur in quantum field theory as Fourier transforms of quantities with support in Γ^+, and so can be analytically continued as Laplace transforms, by using Theorem A2.1. In fact, "this passage via Laplace transforms, from tempered distributions vanishing outside a cone to functions holomorphic in a tube, is one of the basic operations of quantum field theory" and, in view of this, "it is very desirable to characterize exactly what class of analytic functions can appear for a given class of distributions and vice-versa" (Wightman [Wi1], p. 310). Now it is a peculiarity of the class of all analytic functions on a domain $\Omega \subset C^n$, $n > 1$, that they may be analytic on a larger domain called the envelope of holomorphy, obtained by intersecting the domains of analyticity of all the functions that are analytic on Ω. This fact is certainly not true for $n = 1$ since, in this case, one can always find a function analytic on Ω and not continuable. In particular, if $n > 1$, and if a function is analytic outside a bounded set, then it can be continued analytically to include that set. It is an important task to compute the envelope of holomorphy for certain problems in field theory. From the brief discussion above it is clear that this task is nontrivial.

When f_ζ is a suitable distribution on E^n, the Cauchy integral representation of Chapter III extends to read

$$f(p) = \frac{1}{(2\pi)^n} \left\langle f_\zeta, 1 / \prod_{j=1}^{n} (p_j - i\zeta_j) \right\rangle, \tag{A2.11}$$

where $\zeta = (\zeta_1, \ldots, \zeta_n)$ and $\operatorname{Re} p_j \neq 0$ (for $n = 1$ this is the Cauchy integral well defined for $f_\zeta \in \mathscr{D}'_{L_2}$). Equation (A2.11) defines 2^n functions in the 2^n domains $\{p|\pm\sigma_j > 0\}$, and one notes that these functions may continue each other (in the case $n = 1$ we have the two half planes of analyticity). Suppose now that

$$\Box = \frac{\partial^2}{\partial x_1^2} - \sum_{i=2}^{n} \frac{\partial^2}{\partial x_i^2}$$

and let $\Box^k$ be the operator $\Box$ iterated k times; for $n = 1$, $\Box^k$ is simply d^{2k}/dt^{2k} in the notation of Chapter III. Then for every slow-growth causal u there exists a function $\lambda(p - i\zeta)$ on E^n, with fixed $p \in T^+$, such that for some integer k one has

$$\Box^k \mathfrak{F}(ue^{-(\sigma,x)}) = \langle \hat{u}_\zeta, \Box^k \lambda(p - i\zeta)\rangle. \qquad (A2.12)$$

This relation is known as the Vladiminov representation, and appears in field theory (see Bremermann [Br1]).

When $n = 1$, $\lambda(p - i\zeta)$ is just $1/p - i\zeta$ times a constant, and we have

$$D^{2k}\mathscr{L}u = \frac{(i)^{2k}}{2\pi}\left\langle \hat{u}, \frac{d^{2k}}{d\zeta^{2k}}1/p - i\zeta\right\rangle, \qquad (A2.13)$$

which is the Cauchy derivative formula (3.9).

Another question of importance in field theory is that of continuation across a boundary. Thus if $u \in S'$ has its support in both the forward and backward cones when can we continue the holomorphic extension of u, which is defined in $T^+ \cup T^-$, across a portion Λ of the boundary in E^n? For $n = 1$ an answer was provided by Theorem 3.14. The next result is an extension.

Theorem A2.2. Let $u_1(p)$ be holomorphic in T^+ and $u_2(p)$ in T^-. Suppose that on $\Lambda \subset E^n$ one has

$$u_1(p) - u_2(p) \overset{\mathscr{D}'}{\to} 0$$

as $\sigma \to 0$. Then there exists a function $g(p)$, holomorphic on $T^+ \cup T^- \cup \Lambda$, which satisfies $g(p) = u_1(p)$ in T^+ and $g(p) = u_2(p)$ in T^-.

Theorem A2.2 appears as the "edge of the wedge" theorem in much of the field theory literature. It will be of interest of sketch through the idea of the proof here, following an argument in Streater and Wightman [ST1]. First one regularizes u_1 and u_2 to form sequences $\{u_{1,n}\}$ and $\{u_{2,n}\}$. The regularized functions are themselves analytic in the same domain (one establishes analyticity in each variable and then uses Hartog's theorem), and take on the same boundary value on Λ in a continuous manner. An extension to C^n of the classical Painleve result then assures us of continuations $g_n(p)$ which equal $u_{1,n}$ and $u_{2,n}$ in their respective domains. The trick now is to show that the functions g_n are themselves regularizations of some distribution $h_{\sigma,\omega}$. This trick requires the use of the Schwartz kernel theorem but we forego details. Then, as $n \to \infty$, it is shown that $g_n(p)$ converges in the $\mathscr{D}'$ topology to $h_{\sigma,\omega}$ and, hence, as can also be verified, the convergence is uniform on every compact subset of the extended domain. But then $h_{\sigma,\omega}$ is analytic on $T^+ \cup T^- \cup \Lambda$, and is the required continuation.

We have seen how some of the basic results of Chapter III find their setting in C^n. The significance of Theorems A2.1 and A2.2 in quantum field theory, among others, is discussed in articles by Wightman [Wi1], Streater and Wightman [ST1], and Bogolybov and Vladimirov [Bo1].

Bibliography

B1   Beltrami, E. J., Pointwise and norm convergence of distributions, *J. Math. Mech.* **14,** 99–108 (1965).

B2   Beltrami, E. J., and M. R. Wohlers, Distributional boundary value theorems and Hilbert transforms, *Arch. Rational Mech. Anal.* **18,** No. 4, 304–309 (1965).

B3   Beltrami, E. J., and M. R. Wohlers, Distributional boundary values of functions holomorphic in a half plane, *J. Math. Mech.* **15,** No. 1, 137–146 (1966).

B4   Beltrami, E. J., and M. R. Wohlers, On a converse to a theorem of Fatou, Abstract 615–4, *Am. Math. Soc.* (1964).

Bo1 Bogolyubov, N. N., and V. S. Vladimirov, On some mathematical problems of quantum field theory, *Proc. Intl. Congr. of Mathematicians* (1958), pp. 19–32. Cambridge Univ. Press, London and New York, 1960.

Br1 Bremermann, H. J., "Distributions, Complex Variables, and Fourier Transforms," Addison-Wesley, Reading, Massachusetts, 1965.

Br2 Bremermann, H. J., R. Oehme, and J. G. Taylor, Proof of dispersion relations in quantized field theories, *Phys. Rev.* **109,** 2178 (1958).

Br3 Bremermann, H. J., and L. Durand, On analytic continuation, multiplication, and Fourier transformations of Schwartz distributions, *J. Mathematical Phys.* **2,** 240–258 (1961).

Br4 Bremermann, H. J., On finite renormalization constants and the multiplication of causal functions in perturbation theory, ONR Report No. 8, Berkeley, California, 1959.

C1   Cauer, W., The Poisson integral for functions with positive real parts, *Bull. Amer. Math. Soc.* **38,** 713–714 (1932).

D1   Dolph, C. L., Positive real resolvents and linear passive Hilbert systems, *Ann. Acad. Scient. Fenn. A.I.* 336/9 (1963).

F1   Friedman, A., "Generalized Functions and Partial Differential Equations," Prentice-Hall, Englewood Cliffs, New Jersey, 1963.

Fa1 Fatou, P., Series trigonometriques et series de Taylor, *Acta. Math.* **30,** 335–400 (1906).

Fr1 Friedrichs, K., On differential operators in Hilbert spaces, *Amer. J. Math.* **61,** 523–544 (1939).

Fr2 Friedrichs, K., The identity of weak and strong extensions of differential operators, *Trans. Amer. Math. Soc.* **55,** 132–151 (1944).

G1 Gelfand, I. M., and G. E. Shilov, Fourier transforms of rapidly increasing functions and questions of uniqueness for the solution of the Cauchy problem, *Uspehi Mat. Nauk (Russian)* **8,** No. 6, 3–54 (1953); also in *Amer. Math. Soc. (Transl.)* **5,** No. 2, 221–274 (1957).

Ga1 Gärding, L., Some trends and problems in linear partial differential equations, *Proc. Intl. Congr. of Mathematicians* (1958), pp. 87–102. Cambridge Univ. Press, London and New York, 1960.

H1 Hadamard, J., "Le Problème de Cauchy et les Equations aux Derivées Partièlles Lineaires Hyperboliques." Hermann, Paris, 1932.

Hi1 Hille, E., and J. D. Tamarkin, On the absolute integrability of Fourier transforms, *Fund. Math.* **25,** 329–352 (1935).

Ho1 Hörmander, L., Linear partial differential operators. Springer-Verlag, 1963.

K1 König, H., and A. Zemanian, Necessary and sufficient conditions for a matrix distribution to have a positive-real Laplace transform, *SIAM J.* **13,** No. 4, 1036–1040 (1965).

L1 Lauwerier, H. A., The Hilbert problem for generalized functions, *Arch. Rational Mech. Anal.* **13,** No. 2, 157–166 (1963).

La1 Lax, P. D., Theory of functions of a real variable, Lecture Notes, New York University, New York, 1959.

M1 Muskhelishvili, N. I., "Singular Integral Equations" P. Noordhoff, 1953.

N1 Nevanlinna, R., "Eindeutige analytische Funktionen," Springer-Verlag, 1936.

Ne1 Newcomb, R. W., The foundations of network theory, *Elec. Mech. Engrs. Trans. Inst. Engrs. Australia* EM **6,** No. 1, 7–12 (1964).

No1 Noble, B., "Methods Based on the Wiener–Hopf Technique" Macmillan (Pergamon), New York, 1958.

P1 Paley, R. E. A. C., and N. Wiener, Fourier transforms in the complex domain, *Amer. Math. Soc. Collog. Pub.* 1934.

Pa1 Painleve, P., Sur le lignes singulières des fonctions analytiques, *Amer. Soc. Toulouse* **2,** B27 (1888).

Sc1 Schwartz, L., "Theorie des Distributions," vols. I and II. Hermann, Paris, 1957–59.

Sc2 Schwartz, L., Transformations de Laplace des distributions, *Sem. Math. Univ. Lund.* Suppl. 7, Volume M. Riesz, 196–206 (1952).

Sc3 Schwartz, L., Some applications of the theory of distributions, "Lectures on Modern Math," Vol. I. John Wiley, London and New York, 1963.

So1 Sobolev, S. L., Sur un theorem de l'analyse fonctionnelle, *Mat. Sb.* **4** (46), 471–496 (1938).

St1 Streater, R. F., and A. S. Wightman, "PCT, Spin and Statistics, and All That." Benjamin, New York, 1964.

T1 Titchmarsch, E. C., "Introduction to the Theory of Fourier Integrals," 2nd ed. Oxford Univ. Press, London and New York, 1948.

T2 Titchmarsch, E. C., "The Theory of Functions," 2nd ed. Oxford Univ. Press, London and New York, 1939.

Ta1 Taylor, J. G., Dispersion relations and Schwartz distributions, *Ann. Phys.* **5,** 391–398 (1958).

Ti1 Tillman, H. G., Remarks on the multiplication of distributions, Abstract 598–22, *Am. Math. Soc.* (1963).

To1 Toll, J. S., Causality and the dispersion relation: logical foundations, *Phys. Rev.* **104,** 1760–1770 (1956).

W1 Widder, D., "The Laplace Transform," Princeton Univ. Press, Princeton, New Jersey, 1946.

Wi1 Wightman, A. S., Analytic functions of several complex variables, *in* "Dispersion Relations and Elementary Particles," Wiley, London and New York, 1960.

Wo1 Wohlers, M. R., and E. J. Beltrami, Distribution theory as the basis of generalized passive-network analysis, *IEEE Trans. Circuit Theory* **CT-12,** 164–169 (1965).

Wo2 Wohlers, M. R., A distributional study of the real frequency behavior of passive systems, Proc. Allerton Conference, Univ. Illinois, 1965.

Y1 Yosida, K., "Functional Analysis," Springer-Verlag, 1965.

Yo1 Youla, D. C., L. J. Castriota, and H. J. Carlin, Bounded real scattering matrices and the foundations of linear passive network theory, *IRE Trans. Circuit Theory* **CT-6,** 102–124 (1959).

Yo2 Youla, D. C., Representation theory of linear passive networks, MRI Report No. R–655–58, Polytechnical Institute of Brooklyn, 1958.

Z1 Zemanian, A., "Distribution Theory and Transform Analysis," McGraw-Hill, New York, 1965.

Z2 Zemanian, A., An N-port realizability theory based on the theory of distributions, *IEEE Trans. Circuit Theory* **CT-10,** 265–274 (1963).

INDEX